Biological Membranes

TERTIARY LEVEL BIOLOGY

A series covering selected areas of biology at advanced
undergraduate level. While designed specifically for course
options at this level within Universities and
Polytechnics, the series will be of great value to
specialists and research workers in other fields who require
a knowledge of the essentials of a subject.

Titles in the series:

Biological Membranes	Harrison and Lunt
Water and Plants	Meidner and Sheriff
Comparative Immunobiology	Manning and Turner
Methods in Experimental Biology	Ralph
Experimentation in Biology	Ridgman
Visceral Muscle	Huddart and Hunt
An Introduction to Biological Rhythms	Saunders
Biology of Nematodes	Croll and Matthews
Biology of Ageing	Lamb
Biology of Reproduction	Hogarth
An Introduction to Marine Science	Meadows and Campbell
Biology of Fresh Waters	Maitland
An Introduction to Developmental Biology	Ede
Neurosecretion	Maddrell and Nordmann
Physiology of Parasites	Chappell
Biology of Communication	Lewis and Gower

TERTIARY LEVEL BIOLOGY

Biological Membranes

Their Structure and Function

Second Edition

ROGER HARRISON, M.A., Ph.D.

Senior Lecturer in Biochemistry
School of Biological Sciences
University of Bath

GEORGE G. LUNT, B.Sc., Ph.D.

Senior Lecturer in Biochemistry
School of Biological Sciences
University of Bath

Blackie
Glasgow and London

Blackie & Son Limited
Bishopbriggs
Glasgow G64 2NZ

Furnival House
14–18 High Holborn
London WC1V 6BX

International Standard Book Numbers
Paperback 0 216 90998 8
Hardback 0 216 90997 X

Printed in Great Britain by
Thomson Litho Ltd, East Kilbride, Scotland

Preface
to the Second Edition

RESEARCH INTO MEMBRANE-ASSOCIATED PHENOMENA HAS EXPANDED VERY greatly in the five years that have elapsed since the first edition of *Biological Membranes* was published. It is to take account of rapid advances in the field that we have written the present edition.

There is now general acceptance of the fluid mosaic model of membrane structure and of the chemiosmotic interpretation of energetic processes, and our attention has shifted from justifying these ideas to explaining membrane functions in their terms. Much more information has become available concerning the role of the plasma membrane in the cell's recognition of and response to external signals, and this is reflected in the increased coverage of these topics in the book.

The general form of the book remains the same. As before, a list of suggested reading, sub-divided by chapter, is provided and this has been expanded to include a greater proportion of original papers. The book is still primarily designed as an advanced undergraduate text and also to serve as an introduction for post-graduate workers entering the field of membrane research.

We have taken cognizance of the comments of many reviewers, colleagues and students on the first edition and thank them for their contributions. In particular we wish to acknowledge our colleagues R. Eisenthal, G. D. Holman, D. W. Hough, and A. H. Rose. Dr. C. R. Palmer and his staff in the Education Services Unit at Bath University gave great help with the diagrams. Especial thanks are due to June Harrison and Pat Waller for their efficient typing of the manuscript.

Finally we thank the Publishers, Blackie & Son, for their instigation of the first edition and for their continued enthusiasm, support and restraint throughout the subsequent development of the book.

R. HARRISON
G. G. LUNT

Contents

CHAPTER ONE

THE CELL

ALL STUDENTS OF BIOLOGY ARE FAMILIAR WITH THE OBSERVATIONS OF THE English microscopist Robert Hooke. In an account of his investigations published in 1665, Hooke described the microscopic appearance of cork and used the word *cell* to identify the small box-like compartments that he saw. Some 10 years later Antoni van Leeuwenhoek published numerous drawings of bacteria, spermatozoa, protozoa and red blood cells. It was not until the early nineteenth century, however—over 100 years after the observations of Hooke and van Leeuwenhoek—that the anatomist Theodor Schwann announced that 'the elemental parts of tissues are cells, similar in general but diverse in form and function'. Other scientists quickly accepted the new cellular theory, and the scientific world witnessed the beginnings of an explosion in cell science.

The cell membrane

At about the time of Schwann's announcement of the cellular organization of tissues, new microscopes were being developed with compound lenses. These were free from the chromatic aberration that had always bedevilled users of earlier instruments. With the widespread use of the new instruments, and the parallel development of staining procedures for improving the contrast of the various parts of tissues, it became apparent that within the cell, in addition to the nucleus, there was a whole range of hitherto unstudied bodies. Thus in the late nineteenth century mitochondria, Golgi bodies, and a variety of vacuoles and granules, were seen and characterized morphologically. However, in spite of rapid improvements in staining and sectioning, and in the design of lenses and microscopes, no information emerged on the nature of the cell boundary.

In 1855 Karl Nägeli observed differences in the penetration of pigments into damaged and undamaged plant cells, and examined the cell boundary, to which he gave the name *plasma membrane*. He also carried

1

out experiments in conjunction with Cramer which showed cells to be osmotically sensitive, i.e. cells could change their volume depending on the osmotic strength of the surrounding medium. Nägeli suggested that it was the cell boundary that was responsible for these osmotic properties.

Information now began to appear on the function of the still unseen cell boundary or membrane. The German botanist Wilhelm Pfeffer carried out numerous experiments on the osmotic behaviour of plant cells, and concluded that it was the plasma membrane that regulated the uptake of substances into cells. Furthermore he suggested that the barrier was only a few molecules thick. Other workers also examined the selectivity of the plasma membrane and in 1899 Charles Overton published the results of his observations on the entry into cells of a wide range of compounds. He found that the more polar a molecule, the lower the rate of entry into the cell, whereas the addition of non-polar groups such as alkyl chains increased the rate of entry. Occasional marked deviations from the rule also suggested the presence of what we now term active transport systems (p. 174). Overton concluded that the cell membrane, which controlled the rate of entry of substances into the cell, was lipoidal in nature, and that cholesterol and other lipids were involved. Thus in 1899 the foundations for the idea of a *lipid membrane* surrounding the cell were laid. In 1925, Gorter and Grendel suggested that the lipid was in the form of a bilayer. This arrangement was later to be proposed by Danielli and Davson, and to form the basis of a model of membrane structure which is accepted today (see chapter 6).

Yet Gorter, Grendel, Danielli and all the membrane workers who had gone before had never seen the cell membrane. Despite the developments in this period in the fixing and staining of tissue specimens, and even with the development of phase contrast and interference microscopy, the cell membrane was still unresolved, and it remained unresolved until the advent of electron microscopy. The best light microscopes cannot resolve structures separated by less than half the wavelength of light (about 250 nm, using white light). An ordinary electron microscope has a limit of resolution of about 1 nm.

Figure 1.1. An electron micrograph of a longitudinal section of a prokaryotic cell, the Gram-positive bacterium *Streptococcus faecalis*. The thick cell wall CW, with the underlying plasma membrane PM, is clearly visible. The membraneous invaginations which form the mesosomes M are the only intracellular organelles. The cells were prefixed in 3 % glutaraldehyde and counterfixed in 1 % osmium tetroxide. From M. L. Higgins and L. Daneo-Moore, *J. Cell. Biol.* (1974), **61**, 288–300. Original micrograph by courtesy of Drs. M. L. Higgins and L. Daneo-Moore, Temple University School of Medicine.

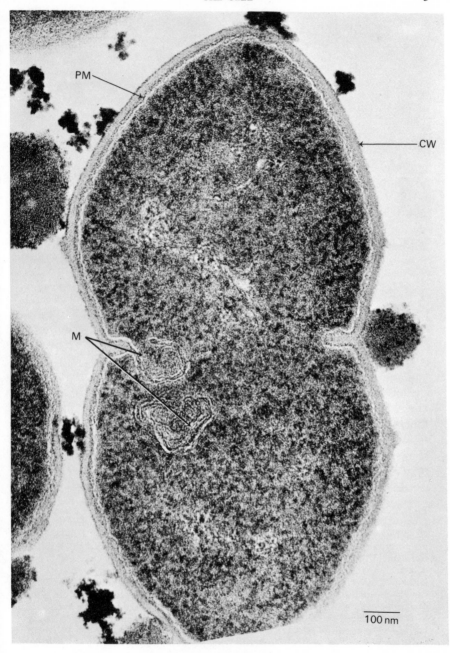

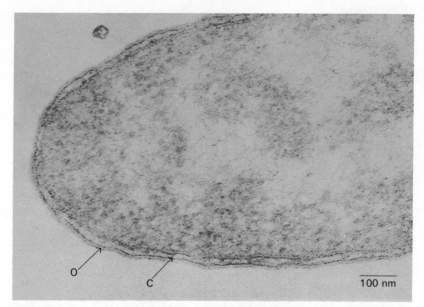

Figure 1.2. Electron micrograph of a section of the Gram-negative marine pseudomonad B16 showing the cytoplasmic membrane C and the outer membrane O separated by the periplasmic region. From J. W. Costerton, J. M. Ingram and K-J. Cheng, *Bacteriol. Rev.* (1974), **38**, 87–110. Original micrograph by courtesy of Dr. J. W. Costerton, The University of Calgary.

Electron microscopy was developing during the period 1930 to 1950 and, when the early electron microscopists turned their attention to cells, there for the first time they saw the boundary membrane of the cell as a solid structural entity. But the animal cells that they first examined seemed to be full of membranes. Far from being a membrane-bound sack with a nucleus and some mitochondria floating in a protoplasmic jelly, the cell appeared to be a membrane-bound collection of membranes. In

Figure 1.3. An electron micrograph of a plasma cell from the submucosa of the rat. The tissue has been fixed in glutaraldehyde followed by osmium. This procedure leads to clumping of the dense granular chromatin C at the centre of the nucleus N, and around its periphery. The less-dense euchromatin extends to the pores P in the nuclear envelope NE (see figure 3.2). The well-developed endoplasmic reticulum ER, which in this cell is predominantly of the rough type, is clearly seen. Free ribosomes R are also present. The stacks of flattened membraneous discs and associated vesicles which constitute the Golgi complex G are readily recognized, as are the mitochondria M. Surrounding the cell is the plasma membrane PM, which, at this magnification, appears as a single dense line. From K. R. Porter and M. A. Bonneville, *Fine Structure of Cells and Tissues*, 1973, Philadelphia, Lea and Febiger, 4th ed. Original micrograph by courtesy of Professor Keith R. Porter, Department of Molecular, Cellular and Developmental Biology, University of Colorado.

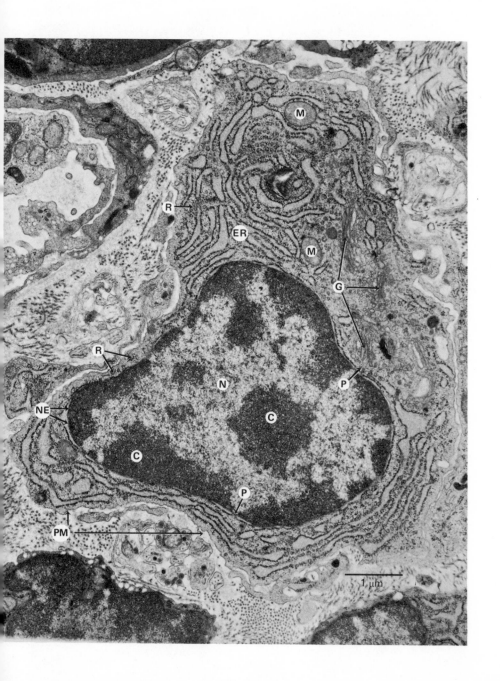

1 μm

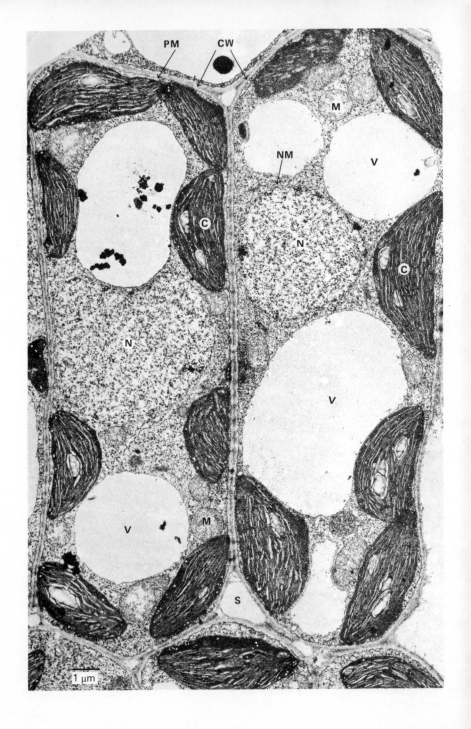

1 µm

the early 1950s, Palade and Porter termed this extensive intracellular membrane system the *endoplasmic reticulum*. Rather surprisingly, all the various membranes within the cell looked very much alike to the early electron microscopists, and all seemed to have the now familiar 'tramline' or 'railroad track' appearance of the plasma membrane (see chapters 3 and 6).

The cell and its membraneous contents

The profusion of intracellular membranes described above is a feature of the cells of higher organisms (plants and animals) and of protozoa, fungi and some algae. Such cells are said to be *eukaryotic*. The bacteria, blue-green algae, and several other micro-organisms are *prokaryotic* cells and have little or no membraneous structure within the cell body. Figure 1.1 shows an electron micrograph of a prokaryotic cell. The cell membrane may be seen clearly with its thick exterior coat or cell wall. Within the cell there is little visible structure; there is for example, no endoplasmic reticulum, no mitochondrion and no discrete nucleus. Many of the functions of these organelles are taken over by the plasma membrane, which has associated with it the electron transport chain and the enzymes of oxidative phosphorylation. The synthesis of many of the molecules required by the cell also takes place at the plasma membrane surface, which may have ribosomes attached to it. The plasma membranes of some bacteria may show invaginations giving rise to organelle-like structures, the *mesosomes* (figure 1.1). It has been suggested that synthesis of membrane and cell-wall components may occur at these sites.

The bacterium shown in figure 1.1 is classified as Gram-positive. Gram-negative bacteria present a rather more complex picture (figure 1.2), in that external to the plasma or *cytoplasmic* membrane is a second *outer membrane* which, although morphologically similar to the plasma membrane, has a different composition (see chapter 5, page 89). The space between the two membranes, the *periplasmic region*, contains a

Figure 1.4. A vertical section through a photosynthetic leaf cell of *Betula verrucosa*. The thick cell wall CW, with the underlying plasma membrane PM, is reminiscent of the structure in figure 1.1. Within the cells, the chloroplasts C and the large vacuoles V are particularly apparent. The nucleus N, with its surrounding nuclear membrane NM, and the mitochondria M, are similar to their animal cell counterparts seen in figure 1.3. Intercellular spaces S are seen. The leaves were fixed in 3% glutaraldehyde and post-fixed in 1% osmium tetroxide; sections were stained with uranyl acetate and lead citrate. From J. D. Dodge, *Ann. Bot.* (1970), **34**, 817–824. Original micrograph by courtesy of Dr. J. D. Dodge, Birkbeck College, University of London.

number of hydrolytic enzymes and binding proteins that may have a role in maintaining the structural integrity of the cell.

The blue-green algae, now increasingly referred to as the *cyanobacteria*, constitute a very large and diverse group of photosynthetic prokaryotes. They have a photosynthetic apparatus which shows many structural and functional similarities to the chloroplast of the eukaryotic higher plants, and thus differ markedly from the majority of prokaryotes in having highly developed intracellular membrane systems.

The eukaryotic cells present a yet more complex picture. Figure 1.3 shows an electron micrograph of a plasma cell from the submucosa of the rat with its wealth of intracellular membraneous structures. In figure 1.4 a photosynthetic leaf-cell is seen. Many of the subcellular organelles in plant and animal cells are similar: thus the nucleus, endoplasmic reticulum, mitochondria and Golgi bodies are comparable. The plant cell is characterized by its outer cell wall (a structure similar to that found in the prokaryotes), the large central vacuole, and the chloroplasts—all of which can be seen clearly in figure 1.4. The functions of the subcellular organelles vary widely; and yet early observations of the intracellular membranes, which are the structural elements of the organelles, suggested that they had a common structure. It is now clear, however, that each membrane has a unique composition and structure, both of which subserve its cellular function.

SUMMARY

1. Cellular theory and the existence of a cell boundary were firmly established in the early nineteenth century.

2. The existence of the cell membrane, its permeability characteristics and its lipid-like nature, were all established before the ultrastructure of the membrane was known.

3. The advent of electron microscopy showed the structure of the cell membrane for the first time.

4. Prokaryotic cells contain little or no intracellular membrane, whereas eukaryotic cells contain abundant membraneous structures.

5. The cell membrane and all the intracellular membranes initially appeared to have the same basic structure.

CHAPTER TWO

MEMBRANE FUNCTION

The plasma membrane

The fundamental function of the plasma membrane is that of *protection*. Thus the cell can maintain a constant internal environment, irrespective of changes that may occur outside. However, the cell membrane, while protecting the cell from a variable external environment, must allow selective communication with the exterior. Arrangements must be made for the controlled passage of nutrients into the cell, and the removal of waste products from it. Plasma membranes, therefore, have associated with them a range of *transfer systems* which enable molecules to pass through the membrane in a specific manner. In most prokaryotic cells, no intracellular membranes are present, and the cell interior is a single compartment bounded only by the plasma membrane. Eukaryotic cells, however, have numerous intracellular membrane systems which form a series of intracellular compartments, within which many of the processes of metabolism occur.

Compartmentalization

Some 15 years ago, E. Pollard proposed that any cell with dimensions greater than 1 μm must have within it regions of submicroscopic order, so that interacting components may approach sufficiently closely for reactions to proceed. Many prokaryotic cells have dimensions of this order, whereas eukaryotic cells are invariably larger. The intracellular membranes of the eukaryotic cells provide the means for ensuring high local concentrations of substrates, by compartmentalizing them within organelles.

The compartmentalization of substrates and products has important implications for metabolic control. The rate of entry of substrate into a compartment and, therefore, the rate of a particular metabolic process,

9

depends on the permeability characteristics of the corresponding boundary membrane. Thus membranes can exert a controlling influence on the metabolic steps proceeding within the organelles. Processes which compete for a common metabolite may be separated by having their enzyme systems in different compartments, and their relative rates controlled by varying the rate of entry of the metabolite. Thus metabolic control is greatly facilitated by a well-developed intracellular membrane system.

Transfer systems

A corollary of the confinement of the cell contents within a plasma membrane and the further compartmentalization of metabolism within the cell is an abundance of membrane transfer systems. These allow for the selective movements of molecules between the cell and its external environment, and between the intracellular compartments. Such transfer systems often mediate undirectional processes and this asymmetry of function is reflected in an asymmetric structure of the membrane (chapter 6).

Most biology students will have seen an amoeba under the light microscope and observed it engulfing water or small particles. This process—*pinocytosis* in the case of liquid, *phagocytosis* if solids are engulfed—is a method for the *bulk transfer* of material into or, if reversed, out of the cell. Transfer of material in this way is probably not dependent on selective changes in the permeability of the membrane, and need not involve the passage of molecules through its molecular structure. The bulk-transfer system is not a characteristic of all cells, whereas the transfer of material at the molecular level is a universal feature of membranes.

Individual molecules may cross membranes in a variety of ways, which may be classified into two groups. There are those processes that do not require a direct supply of energy, *passive transfer*, and those that directly consume energy, *active transfer*. Passive transfer involves the movement of molecules down a concentration gradient across the membrane to a low-concentration region. Active-transfer systems usually move molecules in the opposite direction, i.e. from regions of low concentration to regions of high concentration. Both systems involve selectivity in that some molecules will cross the membrane more easily than others.

Active-transfer systems involve the coupling of a transfer process to an energy source. In several cases an ATPase is involved, and it can be

readily demonstrated that the transfer of many molecules across membranes is dependent on a supply of ATP. Classically this was demonstrated in the giant axon of the squid. These cellular extensions are
sufficiently large to enable material to be injected into the cell interior
without damaging the cell membrane. Using this preparation, it was
possible to demonstrate that in axons that have been treated with cyanide
in order to inhibit ATP synthesis, the extrusion of sodium ions is directly
dependent on intracellular injections of ATP.

Both passive and active-transfer systems may involve the close association of the transported molecules with carrier systems having specific
binding sites located at the membrane surface. The complexed molecule
moves through the membrane and is released at the other surface. The
molecular nature of the transport systems and the way in which they may
be linked to an energy source is becoming clearer. The latest views on
these processes are discussed in detail in chapter 8.

The cell membrane must allow selective communication between the
contents of the cell and its exterior, by mediating the transfer not only
of individual molecules but also of information. Thus, in many cases,
mechanisms exist for the reception of messenger molecules, such as
neurotransmitters and hormones, and for the transmission of their
information into the cell interior. These mechanisms are necessarily
associated with the plasma membrane, and will be discussed in chapter 9.

Electrochemical gradients

The cytoplasm of all cells contains many ions and charged molecules, as
does the aqueous medium which constitutes the external environment of
most cells. When we consider the bulk phase of such solutions we know
that the number of positive charges must equal the number of negative
charges. At the submicroscopic level, however, this principle of electroneutrality may be violated and membranes can separate charges, so
producing an electrical imbalance across themselves. Charge (in the form
of ions) will try to move across the membrane to re-establish neutrality,
but membranes are generally rather impermeable to ions and thus the
flow of ions (the *current*) meets with a high *resistance*. Thus membranes
are able to maintain a considerable electrical imbalance or *membrane
potential* $\Delta\psi$.

A membrane potential may be maintained by the active pumping of
ions which also establishes and maintains a chemical concentration
gradient for a particular ion. The net result is an *electrochemical gradient*

of ions which represents a considerable potential energy source for the cell. We now know that such electrochemical gradients constitute a major source of energy for processes such as solute transport and ATP synthesis. Thus many bacteria utilize an electrochemical gradient of protons as their main driving force for transporting solutes across the plasma membrane, whereas intestinal and kidney epithelial cells use a sodium gradient to provide the energy for amino-acid uptake. The utilization of electrochemical gradients as energy sources is discussed in detail in chapter 8.

Membranes as structural supports

Membranes are the structural elements of cells and therefore, in multi-cellular organisms, of the tissues. The non-membraneous constituents of cells form an aqueous phase which of itself is unable to take on and maintain a characteristic form; only by surrounding the aqueous phase with a membrane can constant shape and form be maintained.

Within the last few years it has become clear that there exists in cells a complex structural system, the *cytoskeleton*, concerned with such diverse functions as maintenance of cell shape, cellular mobility, and intracellular transport of macromolecules. The cytoskeleton has a close relationship with membranes, and we are beginning to understand the details of this relationship, particularly with respect to the plasma membrane (chapter 6).

Membranes and enzymes

Beyond this supramolecular level of organization we find that membranes also fulfil an important role as structural supports at the molecular level. Many of the chemical processes of biological systems involve long series of co-ordinated reactions. A is converted to B, which is converted to C, and so on, through many steps. Such sequences will not proceed efficiently if the reactants are simply dissolved in the cytoplasm. Suppose, however, that the various enzymes involved in the sequence are bound in an orderly fashion to the surface of a membrane, as shown in figure 2.1.

The membrane maintains a constant spatial relationship between the enzymes, allowing the product of enzyme 1 to be passed on to enzyme 2, and so on, along a series. In this way the enzyme-catalyzed reaction becomes *vectorial*—a condition which cannot be achieved when the enzymes and other reactants are free in solution. The reaction sequence will proceed more efficiently when organized in a membrane, and numerous examples are to be found, one of the best characterized being the arrangement of the electron transport chain on the inner mito-chondrial membrane (chapter 9). Enzymes that are associated with membranes in this way are said to be *membrane-bound*.

The membrane-bound form of an enzyme frequently shows properties different from those of a soluble preparation. The differences may be expressed as changes in substrate specificity, inhibitor sensitivity, optimum pH, K_m and general stability. The phenomenon is termed *allotopy*, and was first noted with the ATPase of the inner mitochondrial membrane (chapter 9). This enzyme is inhibited by both dicyclohexyl-carbodiimide and oligomycin when attached to the membrane, whereas the solubilized enzyme is unaffected. Furthermore, the soluble enzyme is cold-labile, whereas no such effect is observed with the membrane-bound form. A further example is the enzyme xanthine oxidase, which is associated with the milk-fat-globule membrane (chapter 4). The enzyme is anchored to the membrane by a hydrophobic protein tail, which can be cleaved by tryptic digestion. The cleaved enzyme is still active, but shows a characteristically different substrate specificity which reverts to the original pattern on reassociation with the membrane.

If we consider the great differences in the environments of a soluble enzyme and a membrane-bound enzyme, it is not too surprising that the properties of the enzyme may change. Thus the soluble enzyme is in a

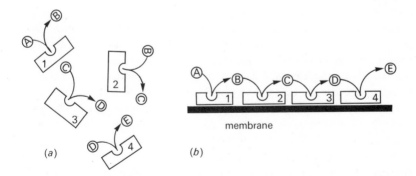

Figure 2.1. Membrane-bound enzymes. In (*a*) enzymes 1–4 and substrates, products A–E are free in solution. When the enzymes are spatially fixed by attachment to a membrane surface, as in (*b*), the conversion of A to E proceeds more readily.

polar hydrophilic environment of high dielectric constant. A variety of molecules including substrates, other metabolites and ions can approach the enzyme easily and may interact with groups on its surface. In contrast, the membrane-bound enzyme may be largely embedded in a lipophilic region of low dielectric constant, with little opportunity for interaction with small polar molecules. The membrane-bound enzyme exists in a relatively stable microenvironment, the nature of which is determined by the molecular arrangement of the membrane. Changes in membrane composition or structure will modify the environment of the enzyme and may constitute a regulatory system.

Cellular recognition

Cells recognize each other. Cellular self-recognition and adhesion is clearly a requirement of organogenesis in higher animals and 'cell-sorting' processes can readily be demonstrated in the formation of homogeneous aggregates from mixed cells of more primitive organisms such as slime moulds and sponges. Moreover, examples of inter-type cell recognition, such as the removal of certain (possibly ageing) erythrocytes by liver cells, are known and have been well studied.

The defence system of the human body involves many instances of cellular recognition processes. Thus leucocytes recognize foreign cells, such as bacteria in the blood, and engulf them, while T and B lymphocytes respond rapidly to foreign or 'non-self' antigens by producing soluble antibodies and by mounting a cellular attack on the invader. The normal exercise of these functions can lead to medical difficulties, such as those of tissue rejection that presently bedevil the transplantation of organs, but when the recognition system goes wrong then more widespread problems can occur. The effectiveness of the immune system depends upon its ability to distinguish 'self' from 'non-self' antigens. Occasionally, however, the system breaks down and 'self' is treated as foreign, often with disastrous results. This happens in a number of *autoimmune disorders* (see chapter 9) and it is clear that a fuller understanding of the recognition processes involved in the immune response is important from a number of medical aspects.

All of these specific cell interactions must depend initially on the presence of particular signals on the plasma membrane of the cells involved. The molecular nature of these membrane markers is currently the subject of a fast-growing research effort and will be discussed in detail in chapter 7.

SUMMARY

1. The plasma membrane ensures the maintenance of a constant internal environment within the cell.

2. The compartmentalization of metabolic processes within membraneous organelles enables high concentrations of reactants to be maintained at discrete sites and provides a means of regulating metabolism.

3. Membranes are selectively permeable, and movement of materials across them usually proceeds in a precisely controlled manner. Such movements may be dependent only on concentration gradients (passive transfer) or may require an energy supply (active transfer).

4. Many membranes maintain large concentration gradients of solutes and are able to couple the potential energy of such electrochemical gradients to a variety of functions.

5. Plasma membranes have an associated cytoskeleton that is involved in cell structure and mobility. Membranes in general also serve as supports and regulators of membrane-bound enzymes.

6. Constituents of the plasma-membrane surface provide a highly sophisticated cellular recognition system.

CHAPTER THREE

MORPHOLOGY OF MEMBRANES

The unit membrane

Early electron micrographs suggested that all membranes have the same basic structure. In the late 1950s and early 1960s, J. D. Robertson proposed his *unit membrane theory* of membrane structure. He suggested that all membranes, whether from plants, animals or micro-organisms have a common structural pattern based on a bimolecular lipid layer with protein associated with the two faces; the pattern, in fact proposed by Danielli and Davson (see chapter 6). Robertson's theory was based on electron-microscope studies of myelin, though in later work he examined a wide variety of membranes and found that they all appeared to have the same structure as myelin.

Unit membranes appeared in transverse section as *trilamellar structures* formed from two electron-dense bands separated by an electron-lucent band. This is the classical 'tramline' picture of early electron micrographs typified in figure 3.1.

Over the years, however, it has become apparent that the original concept of a universal membrane structure is inadequate (see chapter 6). Detailed examination of individual membranes reveals them to have unique structural features which are reflections of their specific functions. Let us now examine in more detail the ultrastructure of some specialized membranes in which a clear correlation between structure and function is emerging.

The nuclear membrane

The nucleus was the first organelle to be clearly recognized using light microscopy and was described by Robert Brown in 1831. It was not until the advent of electron microscopy that the detailed ultrastructure of the boundary separating nuclear material from cytoplasm was elucidated. A

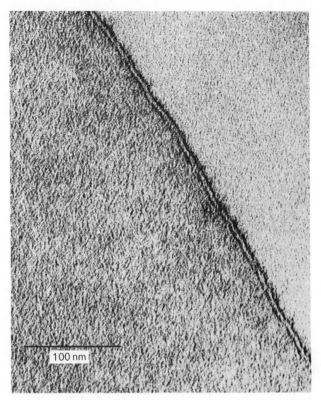

100 nm

Figure 3.1. Electron micrograph of a section of a human erythrocyte fixed in permanganate. The trilamellar unit membrane bounding the cell is clearly defined. From J. D. Robertson in *Cellular Membranes in Development* (1964), (M. Locke, ed.), Academic Press, New York, p. 3. Original micrograph by courtesy of Professor J. David Robertson, Department of Anatomy, Duke University Medical Centre.

major finding was that in eukaryotic cells the nuclear envelope or membrane consists of two concentric layers of membranes. Each layer has the usual tramline appearance, but the membranes are frequently pierced by large pores of between 40 nm and 60 nm diameter. Furthermore, the outer layer of the membrane seems to be continuous with the endoplasmic reticulum, and detailed examination of electron micrographs shows the two membrane systems to be continuous at many points. Most bacteria have no intracellular membranes, and no membrane surrounds the nuclear material. In fungi and cells of higher organisms, the appearance of the nuclear envelope parallels the development of the endoplasmic reticulum. The large pores penetrating the envelope have been studied in detail, and have been found to be very similar in plant and animal cells (see figure 3.2a). The pores have a well-defined structure comprising eight globular or tubular sub-units associated with a system of fibrils and can be isolated from the nuclear envelope as discrete organelles retaining essentially the same morphological appearance as in the intact membrane (see figure 3.2 b and c). Messenger RNA must pass from its intranuclear site of synthesis to the areas of protein synthesis on the ribosomes, and it is attractive to postulate that the role of the pores in the nuclear envelope is to mediate this transfer.

The mitochondrion

Mitochondria are seen in the light microscope as small granular bodies present in all cells of higher organisms. Whereas the contents of the mitochondria are obviously different and separate from the cytoplasm, the nature of the boundary is not at all clear. Electron microscopy reveals the mitochondrion to be a highly-organized membraneous structure. It has an outer membrane surrounding the entire organelle, but within this is another complex membrane system enclosing an inner compartment. Thus the mitochondrion contains two quite separate compartments: an outer chamber or perimitochondrial space, bounded by two different membranes, and an inner chamber bounded by the inner membrane. The inner chamber frequently contains fibrillar and granular material, and is referred to as the *matrix*. Figure 3.3a shows a mitochondrion from a piglet intestinal cell, and the two membrane systems can be clearly seen.

   Examination of electron micrographs reveals that the outer membrane is usually rather thicker than the inner membrane, and frequently shows ordered arrays of small particles. In plant mitochondria, the outer surface of the outer membrane is pitted, and it has been claimed that the pits penetrate to the perimitochondrial space. A redistribution of the particles in the outer membrane has been seen to occur during energization of the electron transport system of the inner membrane, suggesting that

---

Figure 3.2a. Electron micrograph of a platinum replica of a freeze-etched (see page 110) section of a pea-root cell. The inner and outer nuclear membranes, INM and ONM respectively, are seen. The pores P distributed over the nuclear surface, are particularly apparent and can be seen to pass right through both membranes (arrows). Sheets of endoplasmic reticulum ER encircle the nucleus. From D. H. Northcote, *Brit. Med. Bull.* (1968), **24**, 107–112. Original micrograph by courtesy of Dr. D. H. Northcote, Department of Biochemistry, University of Cambridge.

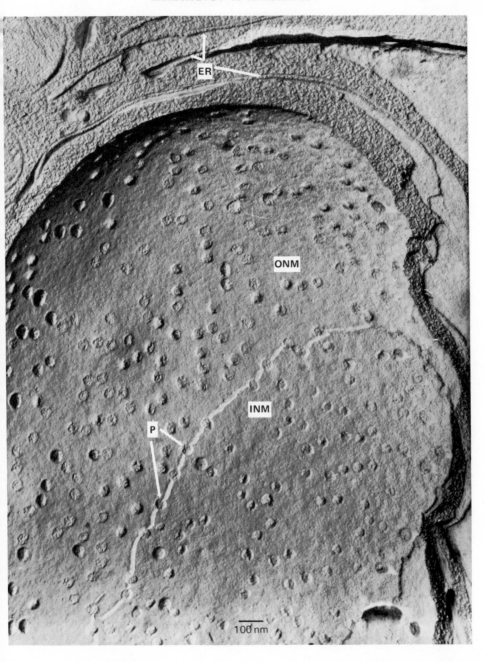

100 nm

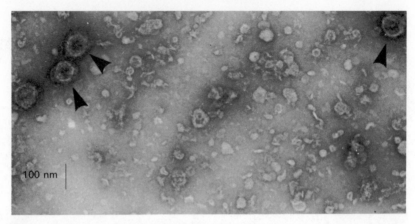

Figure 3.2b. Electron micrograph of a nuclear envelope preparation from rat liver. The material has been negatively stained with ammonium molybdate. The nuclear pores are clearly seen.

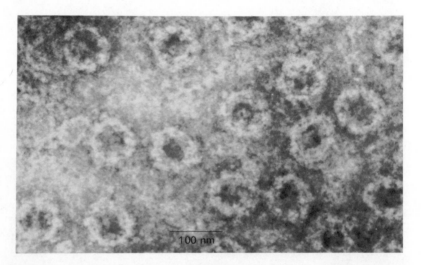

Figure 3.2c. Electron micrograph of rat liver nuclear envelope after ultrasonic disruption of the membrane. As in figure 3.2*b* the preparation has been negatively stained with ammonium molybdate, and free nuclear pore complexes having the same appearance as those in the intact membrane (3.2*b*) are seen.

Both micrographs from J. R. Harris, *Biochim. Biophys. Acta* (1979), **515,** 55–104. Original micrographs by courtesy of Dr. J. R. Harris, North East London Polytechnic.

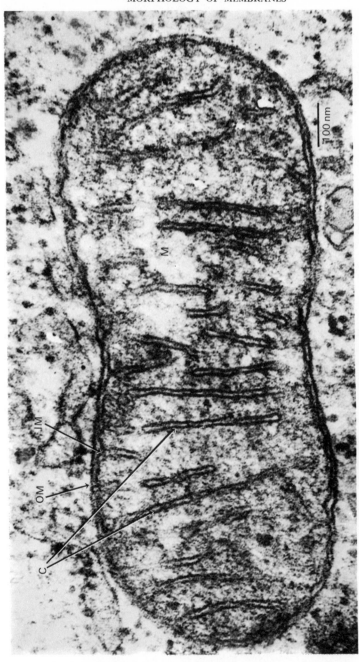

Figure 3.3a. Electron micrograph of a section through a mitochondrion in a piglet intestinal cell. The separate outer and inner membrane systems, OM and IM, are clearly visible. The inner membrane is elaborately folded, giving rise to the cristae C, which penetrate deeply into the matrix M. The tissue was fixed in glutaraldehyde, treated with osmium tetroxide, and stained with uranyl acetate and lead citrate. From E. A. Munn, *The Structure of Mitochondria* (1974), Academic Press, London, p. 3. Original micrograph by courtesy of Dr. E. A. Munn, Agricultural Research Council, Institute of Animal Physiology, Babraham, Cambridge.

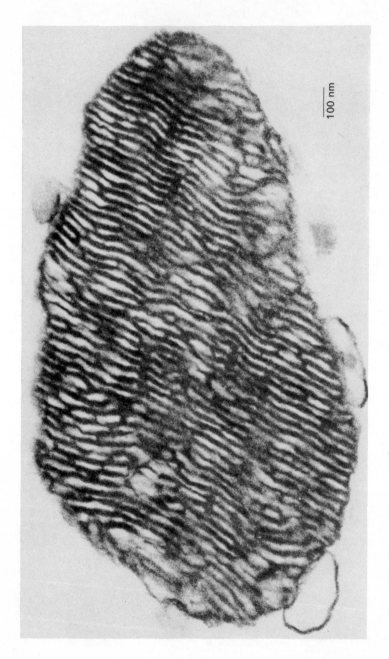

Figure 3.3b. Electron micrograph of a section through a mitochondrion from the flight muscle of a flesh fly. The very extensive inner membrane system is clearly seen. The tissue was fixed with osmium tetroxide and stained with uranyl acetate and lead citrate. Original micrograph by courtesy of Dr. J. F. Donnellan, Shell Research Centre. Sittingbourne.

100 nm

structural-functional changes in the one membrane may be transmitted to the other. The significance of such changes is not yet known.

The inner membrane is a highly convoluted system with numerous infoldings termed *mitochondrial crests* or more usually *cristae* (see figure 3.3*a*). The extent of the folding correlates well with the energy requirements of the tissue. As may be seen in figure 3.3*b*, the mitochondrion from insect flight muscle has a vastly increased inner membrane system, compared with that of the piglet intestinal cell mitochondrion (figure 3.3*a*).

Osmotic rupture of mitochondria under hypotonic conditions, followed by detailed examination of the broken mitochondria in the electron microscope, reveals ultrastructural features unique to the inner membrane. Under these conditions H. Fernandez-Moran showed that the inner surface of the inner membrane is covered with small knobs, 8–10 nm in diameter, linked to the membrane by short stalks. The particles project into the matrix and can be clearly seen in the electron micrograph in figure 3.4.

The biochemical activities of the inner membrane have been shown to be those of *electron transport* and *oxidative phosphorylation*. It is now known that the projections from the inner membrane correspond to the mitochondrial ATPase-ATP synthetase (see chapter 9). The two membrane systems of the mitochondrion have quite different biochemical activities associated with them, but exact correlations between ultrastructure and function have yet to be established.

The chloroplast

Chloroplasts are the most common of the cellular organelles in the plant kingdom. Their average diameter is in the range 4–10 μm, and they are thus readily visible in the light microscope; they are seen to contain numerous small granules termed *grana* embedded in a matrix or stroma. Pigments, such as chlorophyll, can be distinguished in the grana and in the late nineteenth century the role of the chloroplast in photosynthesis was established. Examination of chloroplasts in the electron microscope reveals a complex membraneous substructure. In the higher plants there are two concentric unit-membrane systems, as in the mitochondrion. The outer membrane is a relatively simple structure showing no foldings or invaginations. The inner membrane system, however, is elaborately folded into stacks of flattened cylindrical sacs which form the grana. The individual sacs that collectively make up the grana are referred to as

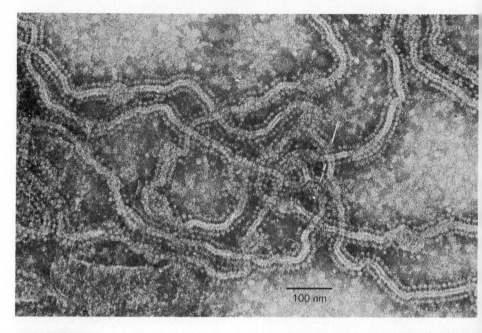

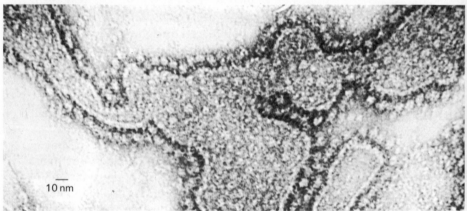

Figure 3.4. Electron micrographs of negatively stained fragments of cristae membranes. The upper micrograph shows a preparation from the flight of muscle *Calliphora* and, in the lower micrograph, rat liver cristae membranes are seen at higher magnification. The stalked particles are clearly visible in both preparations. The upper micrograph is from E. A. Munn, *The Structure of Mitochondria* (1974), Academic Press, London, p. 126. Both original micrographs by courtesy of Dr. E. A. Munn, Agricultural Research Council Institute of Animal Physiology, Babraham.

thylakoid discs. A tubular membrane system, again derived from the inner membrane, is frequently seen connecting the individual grana; this system can be seen in the electron micrograph reproduced in figure 3.5. The same multilammelar system is also seen in the cyanobacteria (page 8) but the boundary membrane is usually absent and therefore no discrete photosynthetic organelle is present. It is possible to disrupt chloroplasts and to recover both complete grana and individual thylakoid discs by a series of centrifugation procedures, when it can be shown that both fractions can carry out the light reactions of photosynthesis. Thus the grana contain both the light-trapping pigments and the enzyme systems involved in the initial light reactions (chapter 9). Attempts to correlate these specific functions of the thylakoid membranes with their structure have met with some success. High-power electron microscopy, particularly of freeze-etched specimens (see chapter 6) has revealed para-crystalline arrays of particles in the thylakoid membrane. These particles are termed *quantosomes* and are about 10–15 nm in diameter, with an estimated weight of some 2 million daltons (see figure 3.6a). The particles occur only in the stacked membrane regions and figure 3.6b shows a freeze-etched spinach thylakoid in which the central stacked membrane is clearly distinguished from surrounding unstacked membrane by closely-packed large particles. It is possible to provoke an unstacking of the membranes by exposure to low-salt solutions in which the particles disperse laterally in the membranes. Restacking *in vitro* can be achieved by adding magnesium ions and/or high-salt concentrations (100–200 mM NaCl). The membranes seem to adhere to each other, and this is followed by a diffusion of the particles into the newly stacked regions. Figure 3.6c shows spinach thylakoids that have been restacked for 1 h and it is clear that the particles are not randomly distributed throughout the membranes.

It has proved possible to isolate quantosomes and to show that they contain pigments, lipids, electron carriers and an ATPase system. Thus the quantosome represents a small self-contained photosynthetic unit. The stacking of the thylakoid membrane restricts the lateral movement of the photosynthetic units, thereby ensuring high local concentrations of light-trapping pigments.

The arrangement of the quantosomes immediately brings to mind the knobs and stalks of the inner mitochondrial membrane and it may well be that similar macromolecular organizations of membrane constituents have evolved in the two organelles to carry out the similar processes of photosynthetic and oxidative phosphorylation (see chapter 9).

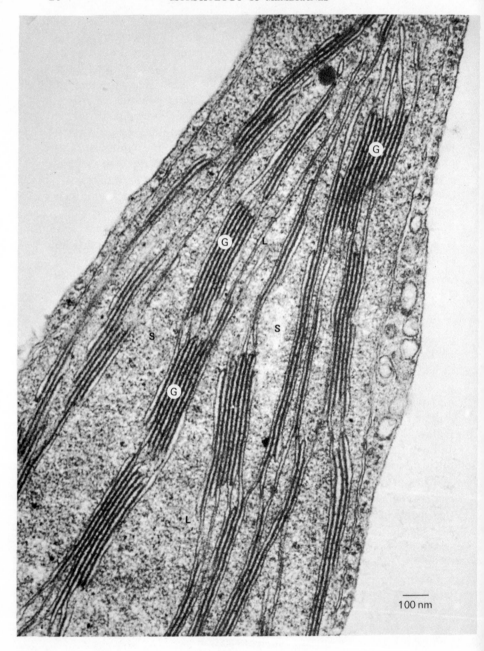

Cell junctions

When we examine the plasma membranes of cells of higher organisms we find that certain regions of the cell surface are highly specialized for cellular contact or communication. These specializations may be broadly classified into three types:

(1) communicating junctions that provide specific pathways for the exchange of materials between adjacent cells,
(2) junctions having a structural role, ensuring adhesion between adjacent cells,
(3) junctions having a sealing function, thereby preventing the free movement of materials through the extracellular spaces between adjacent cells.

The last of these junctional types is particularly common in epithelial tissues where leakage of material through the intercellular spaces would be undesirable.

The three main groups of junctions can be readily differentiated in the electron microscope, and we have a fairly clear picture of the supramolecular organization of the different structures. The communicating junction or *gap junction* forms intercellular channels allowing direct communication between adjacent cells. The permeability of gap junctions is regulated by Ca^{++} ions and it has been demonstrated that injection of Ca^{++} into a cell results in a rapid sealing of these junctions. In the electron microscope the gap junction is seen as a narrowing of the extracellular space to about 10% of its normal width. Intercellular pipes or channels made up of hexameric protein subunits cross the narrowed space and provide for direct communication between neighbouring cells.

The junctions having a role in cell adhesion are called *desmosomes*. In the electron microscope we see the intercellular space filled with fine filamentous structures. Filamentous material is also seen to be associated with the apposed plasma membranes. There is some evidence to suggest that the filaments may be actin-like and able to contract.

The third type of junction is the *tight junction*. At the site of the tight junction the plasma membranes of adjacent cells seem to fuse completely,

Figure 3.5. Electron micrograph of a transverse section through a single chloroplast from a corn marigold leaf cell. The lamellae membranes L are stacked together to form the grana G; single lamellae can be seen crossing the stroma S to connect with other grana. Original micrograph by courtesy of Dr. J. D. Dodge, Birbeck College, University of London.

obliterating the extracellular space. This fusion occurs in a band completely encircling each cell and, as stated above, is particularly prevalent in epithelial cells. Electron-microscope evidence suggests that at the tight junction rows of integral membrane proteins (see page 90) in the two apposed plasma membranes make intimate head-to-head contact— indeed an analogy has been made with the interlocking teeth of a zip fastener—such that the two membranes are locked tightly together. The appearance of the tight junction and of two distinct types of desmosome may be seen in figure 3.7.

The synapse

The *synapse* constitutes a highly specialized junction between the plasma membrane of a nerve cell or neurone and that of another cell. At the synapse a distinct gap called the *synaptic cleft* is seen between the surface membranes of the two cells. We now know that molecules of *neurotransmitter* are released into the cleft and diffuse across it to interact with specific receptors in the membrane of the second cell. This vital link in the nerve pathway has long attracted the attention of histologists, and many ultrastructural features of the membranes have been found that can be related to their unique functions. Figure 3.8 shows a typical synaptic junction in the cerebral cortex of the rat.

Many of the ultrastructural features of the synaptic region were first described by Eduardo De Robertis, and in the mid 1950s De Robertis and Bennett gave the first description of *synaptic vesicles*. These are small unit membrane-bound sacs that are seen exclusively in the pre-synaptic area. It did not take long before the vesicles were shown to be the sites of storage of neurotransmitter, and De Robertis and his co-workers demonstrated that under conditions of prolonged transmitter release the number of vesicles is diminished. The precise mechanism of transmitter release is still the subject of considerable research.

Careful examination of the presynaptic membrane has revealed a unique structural organization. Thus it seems that the presynaptic dense

---

Figure 3.6a. An electron micrograph of a platinum replica of a freeze-etched preparation of a pea-leaf cell. A chloroplast CH is seen lying beneath the inner surface of the plasma membrane Pl. Particulate units (arrowed) are seen on the inner membranes of the chloroplast. The inset shows this region at a higher magnification, and the ordered array of particles is clearly seen. From D. H. Northcote, *Brit. Med. Bull.* (1968), **24**, 107–112. Original micrograph by courtesy of Dr. D. H. Northcote, Department of Biochemistry, University of Cambridge.

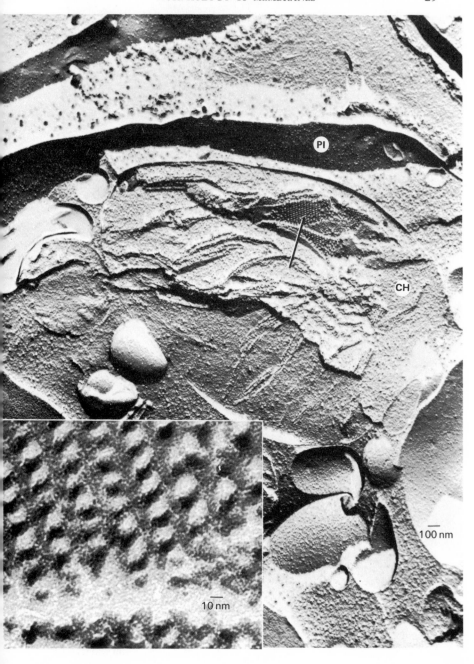

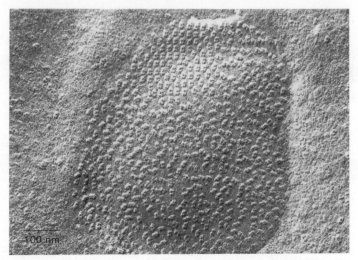

Figure 3.6b. Electron micrograph of the inner surface of spinach thylakoid membrane revealed by deep etching. Closely packed large particles protruding from a smooth background are seen in the central area of the micrograph corresponding to a stacked-membrane region.

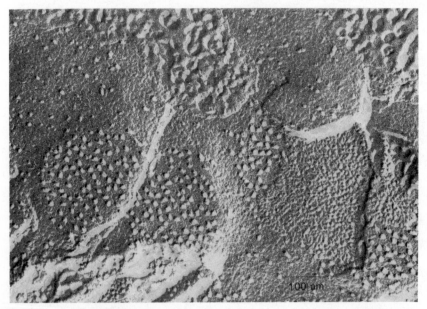

Figure 3.6c. Electron micrograph of spinach thylakoid membranes that have been experimentally unstacked and allowed to restack. The large particles are seen to aggregate in the areas where membrane stacking has occurred. *b* and *c* from L. A. Staehelin, *J. Cell. Biol.* (1976), **71**, 136–158. Original micrographs by courtesy of Dr. L. Andrew Staehelin, University of Colorado, USA.

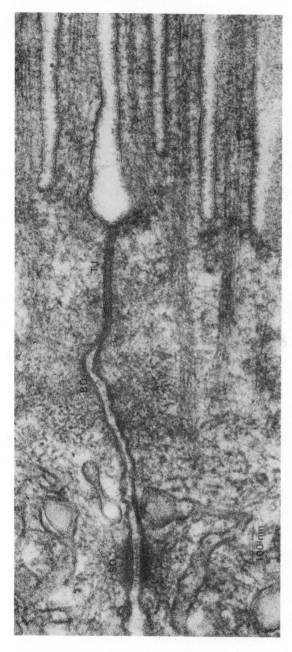

Figure 3.7. Electron micrograph of intercellular junctions in the epithelium of the small intestine. At the far right the adjacent membranes of two cells appear to fuse forming a tight junction TJ. In the centre a band desmosome BD is seen, and dense mats of thin filaments adhering to the inner surfaces of the adjacent membranes. At the far left is a spot desmosome SD. From L. A. Staehelin and B. E. Hull, *Scientific American* (1978) 141–152. Original micrograph taken by Dr. Barbara E. Hull and supplied by courtesy of Dr. L. Andrew Staehelin, University of Colorado, USA.

projections seen in figure 3.8 are part of a complex structure termed the *presynaptic vesicular grid.* Professor K. Akert and his colleagues have suggested that this structure may serve to guide synaptic vesicles down to specific openings or *synaptopores* in the presynaptic membrane. The

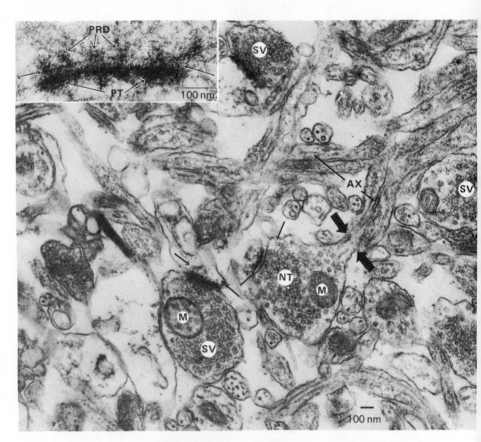

Figure 3.8. Electron micrograph of a section of a slice of rat cerebral cortex. The slice has been incubated at 37°C for 2 hours in Krebs-Ringer buffer solution which has resulted in a considerable enlargement of the extracellular space; the nerve terminals NT are particularly well defined. The terminals are filled with synaptic vesicles SV, and contain some mitochondria M. The synaptic junctions are arrowed. The detailed structure of the junction is seen in the inset enlargement. The synaptic cleft is arrowed and the presynaptic dense projections PRD and postsynaptic thickenings PT are clearly seen. Homogenization of the tissue breaks the axons AX at the point shown by heavy arrows, and the detached terminal may be isolated as a discrete fraction. From G. G. Lunt and E. G. Lapetina (1970) *Brain Res.*, **18**, 451–458.

probable arrangement of the membrane specializations can be seen in
figure 3.9.

Many years ago Sir Bernard Katz showed that the release of neurotransmitter is a *quantal* process, *i.e.* discrete amounts of transmitter
are released into the synaptic cleft. It has always been an attractive idea
that the quanta of neurotransmitter correspond to the synaptic vesicles,

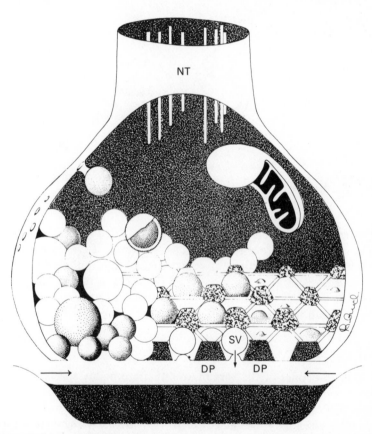

Figure 3.9. Schematic diagram of a presynaptic nerve terminal NT, with presynaptic dense
projections DP, forming the vesicular grid with synaptopores through which synaptic
vesicles SV could discharge their contents (arrow) into the synaptic cleft (arrowed).
Compare with the micrograph in figure 3.8. From K. Akert *et al.* (1972). Freeze-etching
and cytochemistry of vesicles and membrane complexes in synapses of the central nervous
system in *Structure and Function of Synapses* (ed. G. D. Pappas and D. P. Purpura) Raven
Press, New York, pp. 67–86. Original diagram kindly supplied by Professor K. Akert,
Brain Research Institute, University of Zurich.

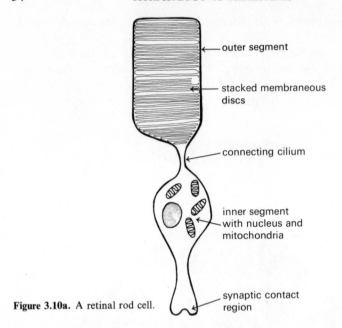

Figure 3.10a. A retinal rod cell.

and when the latter were described by De Robertis, numerous workers sought unequivocal proof that this is so. Although it is still not proven that a direct correlation exists between quanta of transmitter and synaptic vesicles, it is likely that the structural organization of the presynaptic membrane provides a mechanism for channelling single vesicles to sites where they can release their contents into the synaptic cleft.

The retinal sacs

The rods and cones of the vertebrate retina are highly specialized cells which detect light energy and convert it into an electrical impulse. Figures 3.10 *a* and *b* show the main features of a retinal rod cell. The cone cell is essentially the same, except that the outer segment is short and compact.

The outer segment of the cells is concerned with the detection of light, and examination of this part of the cell with the polarization light microscope suggests the presence of a multilamellar structure not unlike that seen in myelin. Electron microscopy reveals that the pattern seen in the polarization microscope results from membraneous sacs stacked

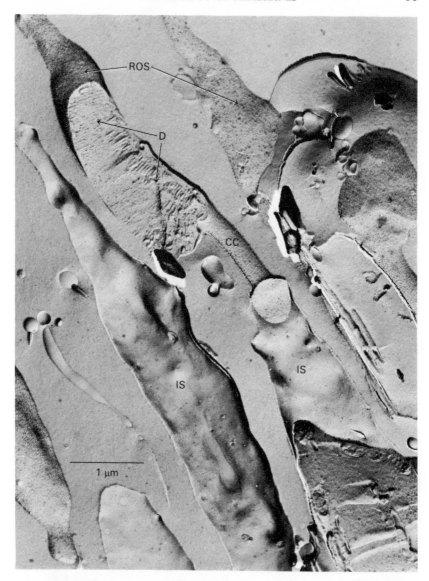

Figure 3.10b. Electron micrograph of freeze-fractured mouse retina showing portions of the rod cell inner segment IS, the connecting cilium cc, and the rod outer segment ROS. The outer segment contains the stacked membraneous discs which are seen in cross-fracture D. From L. Y. Jan and J.-P. Revel, *J. Cell. Biol.* (1974), **62**, 257–273. Original micrograph by courtesy of Dr. Lily Yeh Jan, California Institute of Biology, Pasadena.

longitudinally along the segment axis. The sacs are formed from two-unit membranes which surround a space 3–5 nm wide (figure 3.11). These sacs can be shown to be osmotically sensitive and will swell under hypotonic conditions.

It has been demonstrated that the retinal sacs are derived from the plasma membrane of the cell. In the cones of the frog retina, the plasma membrane undergoes repeated invaginations which lead to the build-up of a series of disc-like structures. In the rods the discs develop in a similar manner, i.e. by the invagination of the plasma membrane, but the resultant sacs seem to pinch off producing stacks of discs completely separated from the bounding plasma membrane (see figure 3.11). It has been shown that the discs contain the visual pigment rhodopsin, made up from a protein, opsin, and a chromophore, 11-*cis* retinal (chapter 9). It is suggested that the rhodopsin is synthesized in the inner portion of

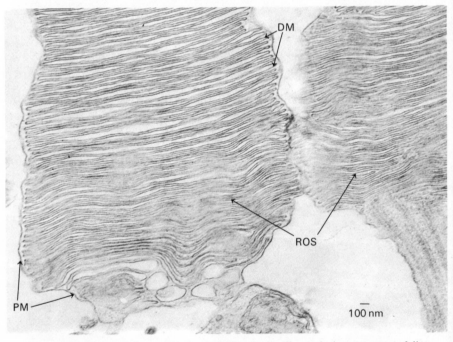

Figure 3.11. Electron micrograph of rat retinal rod cells. The stacked arrangement of discs within the rod outer segment ROS is clearly illustrated, and the disc membranes DM are seen to be quite separate from the plasma membrane PM of the cell. Original micrograph by courtesy of Dr. E. De Robertis, Instituto de Biologia Celular, Universidad de Buenos Aires.

the cell and is then transported, via the connecting cilium, to the base of the outer segment, where it is incorporated into developing retinal discs. The protein is a major structural element of the disc membrane and may comprise 80% of the total membrane protein. There is good evidence

Figure 3.12. Electron micrograph of freeze-fractured retinal disc membranes from the mouse. The stacked arrangement of the discs is clearly seen. The cytoplasmic leaflet of the disc membrane CL contains numerous particles which are thought to represent the sites of rhodopsin molecules. The intra-disc leaflet IL of the disc membrane shows no such particles, which suggested that rhodopsin was located in the outer cytoplasmic half of the disc membrane bilayer (but see page 212). From L. Y. Jan and J.-P. Revel, *J. Cell. Biol.* (1974), **62**, 257–273. Original micrograph by courtesy of Dr. Lily Yeh Jan, California Institute of Biology, Pasadena.

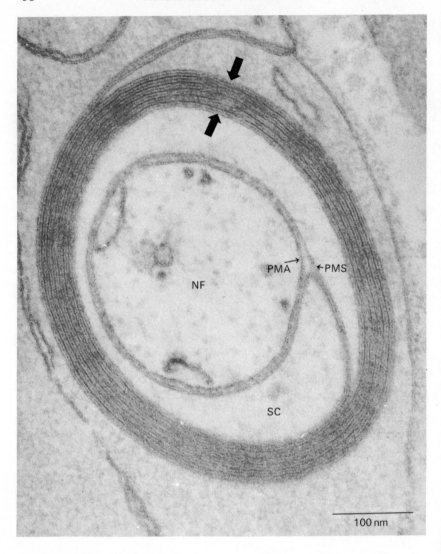

Figure 3.13. Electron micrograph of a young mouse sciatic fibre with a developing myelin sheath. The nerve fibre NF is enveloped in the multilaminar myelin sheath (heavy arrows). The separate plasma membranes of the axon PMA and of the Schwann cell (SC), PMS, can be clearly distinguished. From *Neurosciences Res. Prog. Bull.* (1971), **9**, p. 520, (ed. L. C. Mokrasch, R. S. Bear and F. O. Schmitt). Original micrograph by courtesy of Professor J. David Robertson, Department of Anatomy, Duke University Medical Centre.

that the light-induced isomerization of the chromophore, 11-*cis* retinal, induces a conformational change in the protein of the disc membrane, resulting in a permeability change which releases Ca^{++} from within the disc. The change in concentration of free Ca^{++} within the rod modulates the electrical activity of the rod cell (chapter 9).

The arrangement of flat sacs with light-trapping pigments forming the major component of the membrane is reminiscent of the grana of chloroplasts, and freeze-etched preparations of both membranes show a similar granular appearance (see figure 3.12). Both the retina and the chloroplast are concerned with absorbing the maximum amount of light, and conversion of the light energy to some other form. The multilayered membranes provide a system which is able to do this most effectively.

Myelin

Much of our knowledge of membrane ultrastructure has been obtained from studies on myelin. In 1936 W. J. Schmidt produced models of the structure of myelin membranes based on studies with the polarization light microscope. The structure proposed by Schmidt was very similar to the contemporary model of Danielli and Davson for the plasma membrane. Later studies of myelin using X-ray diffraction confirmed the early prediction of Schmidt, and in 1954 Geren published electron micrographs of myelin that clearly showed the myelin membrane to be continuous with the plasma membrane of the Schwann cell. It is now well known that the myelin sheath which surrounds axons is produced by the Schwann cells, in the case of peripheral nerves, and by oligodendroglial cells in central nerves. The process involves a gradual spiralling of the myelin-producing cell around the axon, thereby producing the characteristic multilayered sheath shown in the electron micrograph of figure 3.13.

Electron microscopy of myelin reveals it to approach a model unit membrane: its main function is to act as an insulator for the axons, a purpose for which its multilayered lipid-protein structure of low dielectric constant makes it ideally suited.

SUMMARY

1. Early electron microscopy of membranes led to the formulation of the *unit membrane* theory. Unit membranes were characterized by their tramline appearance—two electron-dense bands bounding an electron-lucent zone.

2. As techniques improved, many examples of membranes having unique ultrastructural features were found. These are seen to be closely linked to the characteristic function of the membrane. Thus the *nuclear envelope* has a system of large pores through which RNA is thought to pass. The *thylakoid membranes* of chloroplasts and the *inner mitochondrial membrane* have closely packed projections from the membrane surface that are the sites of ATP synthesis. The light-trapping membranes of the *retinal rod* cell have a stacked appearance similar to that of the thylakoid membranes, and the membranes also have a densely packed array of particles that correspond to the visual pigment rhodopsin. At cell junctions there may be considerable modification of the membrane structure: at *gap junctions* pipes or channels cross the membranes of adjacent cells and provide for intercellular communication; *desmosomes* constitute structural links between adjacent cells, and at *tight junctions* the plasma membranes of adjacent cells fuse completely, providing a barrier to the movement of water and solutes between cells. The *synapse* is a highly specialized junction, and the presynaptic membrane has a grid-like structure that facilitates the attachment of synaptic vesicles and the release of their neurotransmitter content into the synaptic cleft. *Myelin* conforms closely to the original pattern of the unit membrane.

CHAPTER FOUR

MEMBRANE PREPARATIONS

THE ULTRASTRUCTURE OF CELL MEMBRANES CAN BE STUDIED WITHOUT separating them from other cell constituents. Similarly, their permeability properties and many transport phenomena can be investigated using intact cells or even whole tissue. However, before we can hope to understand the functions of individual membranes in molecular terms, we need to know their precise chemical composition. Procedures must therefore be devised which allow the complete separation of one particular membrane system from the many membranes that constitute the cell. Having achieved this, it is then necessary to show that the isolated membrane is not substantially different from the intact membrane *in situ.* What we hope to do is to use our knowledge of the isolated membrane *in vitro* to explain and understand its function and behaviour *in vivo.*

The first isolation of a subcellular organelle was achieved by Friedrich Miescher, who in 1871 isolated nuclei from broken cells by a centrifugation procedure. Later workers improved upon Miescher's methods, and in the late 1930s R. R. Bensley and his colleagues isolated mitochondria. It was Albert Claude, however, who, during the period 1937 to 1945, laid the foundations of present-day subcellular fractionation schemes. The earlier workers had concentrated only on the isolation of a single subcellular organelle, and paid no regard to the many other cell constituents. Claude's approach was quite different in that he introduced the concept of a quantitative analysis of the cell contents. Thus analyses were carried out on all the cell fractions. A final balance sheet was drawn up, thereby allowing an assessment of the contribution made by each fraction to the intact tissue. Claude also introduced the idea of characterizing the fractions by measuring enzymic activities, and thus paved the way for the establishment of *marker enzymes* which are now widely used. Many workers now characterize a subcellular fraction solely on the basis of measurements of enzymic activities, without pausing to think that in the first instance a strict correlation between morphology and enzymic activity had to be established.

41

Even though it is now almost thirty years since Claude developed his tissue fractionation scheme, there are very few isolated membrane fractions that fulfil simple criteria of purity. Let us look at some of the procedures that are used both to isolate and to characterize membranes from a variety of cell types.

Cell disruption

The first step towards isolating a particular membrane is, in most cases, a disruption of the cell. The procedure that is adopted can profoundly affect the nature of the ensuing membrane preparations and must be chosen with care. In the case of soft animal tissues, methods based on mild shear forces are most commonly employed, whereas plant and bacterial cells may require extremely high shear forces coupled with high pressures and extremes of temperature.

The composition of the medium in which the cells are broken must also be considered. The commonest medium is a sucrose solution that is usually slightly hypotonic; 0·25 M sucrose is widely used in liver fractionations. The use of salt solutions frequently causes the aggregation of subcellular particles, although reports have been made that in some tissues, such as spleen, sucrose promotes aggregation. In such cases, salt solutions are the preferred media. The homogenization medium is usually buffered at pH 7·4 and there are many reports that at lower pH-values tissues become more difficult to homogenize.

Low-shear-force methods

The most common method of achieving a low-shear disruption of tissue is to use a Potter-Elvejhem homogenizer. As shown in figure 4.1a this consists of a precision-bore glass tube with a close-fitting motor-driven pestle, originally of Perspex but now usually of Teflon. The tissue is cut into small pieces in the buffer, and is then forced between the rotating pestle and the wall of the tube by raising and lowering the glass tube. Thus the cells are subjected to a shear force and are torn open, releasing the subcellular organelles. The radial clearance between the pestle and the tube wall can vary between 0·05 and 0·5 mm, a value being chosen so that minimal damage occurs to nuclei, mitochondria and other organelles. Even with the widest clearances, about 10% of the mitochondria suffer damage to their outer membranes, and there is evidence that at least 15% of the lysosomes and peroxisomes may be broken. A similar disruption of cells can be achieved by using a Dounce homogenizer (figure 4.1*b*) which has a ground-glass tube with a close-fitting ground-glass plunger; this is operated by hand and is considered to be more gentle than the Potter-Elvejhem homogenizer.

The thick polysaccharide cell wall of the plant cell and the extraneous coat or wall of many micro-organisms render such cells resistant to the low-shear forces generated in this type of homogenizer. Methods are being developed, however, in which the wall is digested away, using broad specificity enzymes, leaving an intact plasma membrane-bounded cell or *protoplast*. This procedure has been successfully applied to plant tissues, yeast and bacteria, and it can be shown that the protoplast retains the permeability characteristics of the original cell. The protoplast may

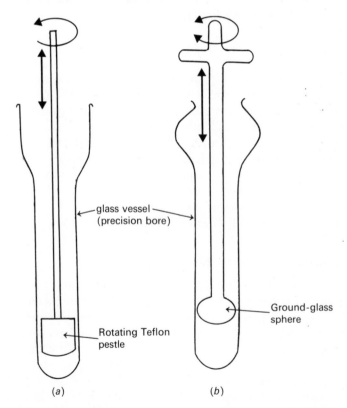

glass vessel
(precision bore)

Rotating Teflon
pestle

Ground-glass
sphere

(a)   (b)

Figure 4.1. (*a*) The Potter-Elvejhem and (*b*) the Dounce tissue homogenizers.

be disrupted in a Potter-Elvejhem homogenizer, thereby yielding membraneous subfractions.

High-shear methods

High-shear forces may be developed by forcing tissue suspensions through extremely small apertures at high flow rates. The French press uses this technique and is often used to break open yeast cells. Preparations of chloroplast membranes have also been described in which the initial cell breakage is achieved in a French press. Further disruption may be achieved by cooling the entire apparatus with liquid nitrogen. Ice crystals within the cells then act as an abrasive agent during the passage of the frozen tissue suspension through the narrow aperture. Such extreme methods are used almost exclusively for the preparation of organelles and membrane fragments from micro-organisms.

Alternative procedures

Many methods of cell disruption other than those based on shear forces have been investigated but have not been widely adopted. *Ultrasonication* of cell suspensions has been used, but this procedure invariably damages the plasma membrane and cell organelles. *Gas-bubble nucleation*, in which cells are equilibrated with inert gas at very high pressures and then suddenly returned to atmospheric pressure, has been used quite successfully in the preparation of plasma membranes from tumour cells, but has not been widely used for other tissues. A major disadvantage of the technique is that the plasma membrane is fragmented into very small pieces which are difficult to separate from microsomes.

Membrane separation methods

Having obtained a cell-free suspension of membraneous fragments, the next step is to separate the particular groups of membranes and to obtain finally a homogeneous population. The most common separations are those based on centrifugation procedures. The suspension is subjected to an artificial gravitational field high enough to overcome the energy of random thermal motion of the particles. So long as the suspending medium is less dense than the particles, they will migrate to the bottom of the centrifuge tube and form a compact pellet. The rate at which individual particles move is determined largely by their size, though shape and density also contribute. With various combinations of increasing gravitational field (*g* force) and time, it is possible to obtain a series of fractions from a tissue homogenate, each fraction being greatly enriched in particular membrane types. For complete purification of a membrane fraction, it is usual to follow this initial *differential centrifugation* with a density gradient step. The most widely used procedure is an *iso-pycnic* (Greek, *same density*) separation in which particles are centrifuged through a density gradient that encompasses their own density. A particle cannot progress beyond this point, as the buoyancy effect from the displaced medium exactly matches the imposed centrifugal field. Thus the particles come to rest at precisely defined zones within the gradient. Fortunately, although the behaviour of most cell membranes in differential centrifugation precludes completely quantitative separation, there are sufficient differences in the densities of the various membraneous fractions from cells to allow excellent separation in the density-gradient centrifugation procedures.

There are occasional reports of quite different methods of separation and purification of membranes. Thus membraneous fractions of brain and of kidney tubule cells have been separated on the basis of their surface charge by electrophoretic procedures. Gel exclusion chromatography and affinity chromatography, the latter particularly in immunological studies, have also been used to isolate different membrane types.

Membrane characterization

Once an isolated membrane fraction has been identified morphologically, other factors may be sought which aid in its characterization. If an enzyme can be shown histochemically to be exclusively associated with a certain membrane fraction, it can be used as a *primary* marker. Although histochemistry has permitted the subcellular localization of a number of enzymes, there are relatively few which lend themselves to exploitation at the level of the electron microscope. However, the rat liver plasma membrane may be identified by the presence of the primary enzyme markers, $5'$ nucleotidase, Na^+,K^+-ATPase and alkaline phosphatase. When the plasma membrane has been characterized in this way, other enzyme activities in the fraction may be used as *secondary* markers to identify the fraction in subsequent preparations. In rat liver plasma membranes, such secondary markers would be leucine aminopeptidase and adenylate cyclase.

Other methods of characterizing the plasma membrane which are not dependent on the presence of a particular enzyme have also been sought. One such approach has been to label the membranes, prior to cell disruption, with a fluorescent or radioactive molecule which is known to interact with membrane constituents, but which will not enter the cell. By following the distribution of the label throughout the fractionation procedure, we will hopefully also be following the plasma membrane. The criticism of this method is that once the membrane is labelled in this way it ceases to be the same structure as it was *in vivo*.

Other characteristics of plasma membranes that have been exploited in an attempt to characterize the isolated fractions are their surface glycoproteins (see chapter 7) and their lipid content. The presence of sialic acid has been used as a plasma membrane marker, but sialic acid has also been shown to be present in other organelles, notably mitochondria, albeit at rather lower concentrations. In general, the plasma membrane has much higher cholesterol levels than other cell organelles, though again this alone is not sufficient for us to be able to assign the title of plasma membrane to a fraction. Unequivocal proof of identity can come only from a number of different analyses. Morphological, histochemical, enzymological and chemical assays should all be carried out, and only after a consideration of the combined results of such complementary analyses can the isolated fraction be unequivocally classified.

Animal cell membranes

The erythrocyte membrane

The mammalian erythrocyte membrane has long been the favoured material for studies on membrane composition and on structure-function relationships. There are several reasons for this: it is much easier to prepare in a pure and relatively intact form than most other membranes and, equally important, it is available cheaply in large quantities. Furthermore the erythrocyte is relatively easy to prepare in an empty state, i.e. the cell contents may be removed leaving an almost unchanged plasma membrane. Such preparations, designated *ghosts*, provide a particularly attractive system for studying the transport properties of the plasma membrane. The mammalian erythrocyte is an anuclear cell containing no organelles, and is in effect a plasma membrane containing only cytoplasm. This atypical situation has led many people to criticize the extensive use of the erythrocyte membrane as a model for plasma membranes. However it should be borne in mind that this membrane

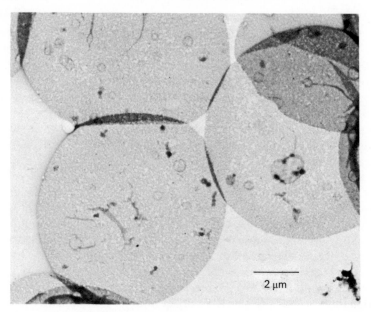

Figure 4.2. Electron micrograph of unstained human erythrocyte ghosts. The ghosts have the appearance of empty sacks with some slight 'crumpling' of the surface. From G. L. Nicolson. *J. Cell. Biol.* (1973), **57**, 373–387. Original micrograph by courtesy of Dr. J. R. Smith, Department of Cancer Biology, The Salk Institute for Biological Studies.

carries out many of the functions of a normal plasma membrane. It contains a wide range of enzymes, shows specific vectorial transport properties, and contains a range of lipids and proteins not greatly different from those of other plasma membrane preparations.

The preparation of the ghost involves removal of the cytoplasm with its high concentration of haemoglobin without unnecessarily damaging the membrane. Many methods have been developed for doing this, each group of workers having their own particular ideas on the best method, but basically all involve a controlled haemolysis of the cell. This is generally a hypotonic (20 mosM) lysis at a pH of about 7·6, the membranes being collected by sedimentation. This procedure may lead to a completely haemoglobin-free membrane which is the subject of considerable controversy. Many workers maintain that *in vivo* some haemoglobin is bound to the membrane, and indeed should be considered as part of the membrane structure. Proponents of the haemoglobin-free membrane preparations answer that some haemoglobin may well be associated with the membrane, but that it is only loosely adsorbed onto the membrane's inner surface, and should not be considered a membrane constituent. We have no way of knowing which view is correct at present, but it seems quite likely that no sharp division between what is and what is not a membrane constituent can be made. Certainly erythrocyte membrane preparations, with or without haemoglobin, show a wide range of enzyme activities. Morphologically they appear to be little changed and when resealed (see chapter 5) retain the osmotic characteristics of the intact erythrocyte. Thus the erythrocyte membrane or ghost as shown in figure 4.2 can be considered as a reasonable model for the natural membrane *in vivo*.

Other plasma membranes

The isolation of plasma membranes from cells other than the erythrocyte presents many problems, most of which have not yet been overcome. The cells of any organ are present as a solid structured mass which must be disrupted before the isolation of the membranes can begin. Organs invariably contain more than one type of cell, and in an ideal situation a cellular fractionation would precede any attempts at a subcellular fractionation. In practice this is hardly ever done, and the heterogeneity of cell types within the organ in question is usually conveniently forgotten. Further contamination of membrane fractions can arise from both the vascular and nervous systems associated with the organ, but again these are usually ignored. In spite of numerous pitfalls that lie in the path of the isolator of plasma membrane, many preparations have been made and characterized. The most popular starting material has been rat liver and, although the liver cell is frequently given as an example of a 'typical' animal cell, it should be borne in mind that liver is made up of roughly equal numbers of parenchymal and reticulo-endothelial cells. Despite this fact, most workers simply disrupt the whole liver tissue.

It is still difficult to characterize absolutely a plasma membrane fraction. The membrane vesiculates during the homogenization and appears similar to other intracellular membranes, though usually the fragments are rather bigger. In the particular case of liver cells there are morphological features which assist in the identification of the isolated membrane. Thus there are characteristic membrane specializations— *junctional complexes* (chapter 3)—which occur at the points where neighbouring cells make contact. These connections survive the homogenization procedure, and pieces of plasma membrane attached to the junctional complex can be recognized in the electron microscope (see

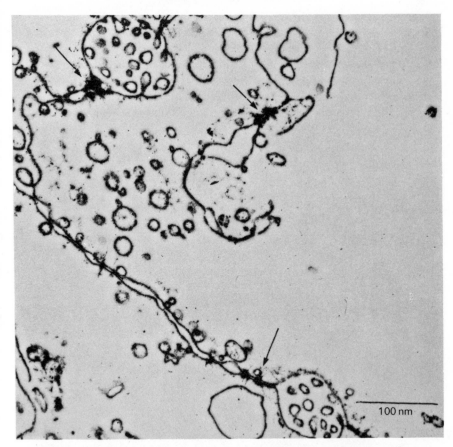

100 nm

Figure 4.3. Electron micrograph of isolated plasma membranes of rat liver cells. The large sheets of plasma membranes from several cells are still attached to each other by way of the junctional complexes (arrowed). From P. Emmelot *et al.*, *Biochim. Biophys. Acta* (1964), **90,** 126–145. Original micrograph by courtesy of Professor P. Emmelot, Antoni Van Leeuwenhoekhuis Het Nederlands Kankerinstitut, Amsterdam.

figure 4.3). There are also reports that after negative staining, hexagonal arrays are visible on the plasma membrane, no such feature being visible on any of the intracellular membranes.

The membrane surrounding fat droplets in fresh milk is derived from the mammary-cell plasma membrane during secretion. Fat globules are formed in the basal region of the mammary cell and migrate to the apical cell surface, through which they bulge, becoming progressively enveloped in the plasma membrane. Finally the protruberance is pinched off, resealing the cell, and leaving an intact membrane around the departing fat globule (see figure 4.4). In some 1–5 % of the globules, small crescents of cytoplasm can be detected trapped between the fat droplet and the surrounding membrane, but very little cytoplasmic material is carried away by the fat globule in this way.

There is evidence that Golgi vesicle membrane is also directly incorporated into milk-fat-globule membrane during the secretion process.

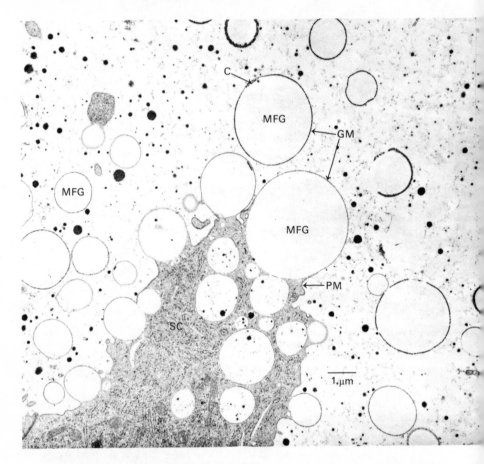

Figure 4.4a. An electron micrograph of a section of goat mammary tissue showing milk fat globules MFG, in the process of secretion. The globules approach the plasma membrane PM of the secretory cell SC, and bulge through it. The globule forms a protuberance, which pinches off, forming a free milk-fat globule with a boundary membrane GM. Small crescents of cytoplasm C are sometimes trapped in the globules.

This does not necessarily imply that milk-fat-globule membrane differs from the apical plasma membrane of the mammary cell, since this plasma membrane is also continually being replenished by Golgi vesicle membranes as the vesicles discharge their contents into the alveolar lumen.

The laboratory preparation of milk-fat-globule membranes from expressed milk is simple. Separated cream is well washed to remove soluble skim-milk components and shaken (the equivalent of churning); the

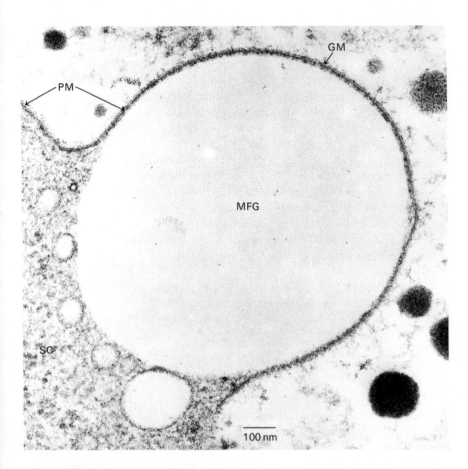

Figure 4.4b. At higher magnification it can be clearly seen that the trilamellar plasma membrane PM of the secretory cell SC is continuous with the globule membrane GM. Both original micrographs kindly supplied by Dr. F. B. P. Wooding, Agricultural Research Council Institute of Animal Physiology, Babraham.

fat-globule membranes rupture, allowing the fat to coagulate as butter, and leaving a suspension of membranes in the buttermilk. In view of the practical difficulties involved in obtaining clean mammalian-cell plasma-membrane preparations, the milk-fat-globule membrane constitutes a particularly convenient large-scale supply of relatively pure plasma-membrane material.

Advances in our ability to maintain higher cells in culture, coupled with the possibility of genetically manipulating such cells, provide a new opportunity to investigate the relationship between plasma membrane structure and function. In essence this approach to the study of the plasma membrane derives from the selection of particular *mutants* in which specific genetic changes give rise to structural and functional modifications in the plasma membrane. In this way cells having, for example, an altered $Na^+,K^+,ATPase$ or an altered adenylate cyclase have been studied and may provide the means for novel studies on these two important plasma membrane enzyme systems (see chapters 8 and 9).

In addition to providing material for fundamental research into membrane structure and function, the techniques may be of great importance in investigating particular diseases. Thus it is possible to culture cells from patients with genetic disorders and to examine the membranes of such cells for altered activities. Some success has already been achieved along these lines in that cells from hypercholesterolaemia patients have been cultured and seen to be defective in internalizing high-density lipoprotein.

The availability of such mutant cell lines is increasing rapidly, and they will provide a powerful new tool to aid our investigations of the functional activities of the structural components of the plasma membranes of mammalian cells.

Intracellular membranes

In general, animal intracellular membranes have not been as well characterized as the plasma membrane; thus many fractionation schemes give rise to fractions enriched in a particular *organelle* which may contain more than one membrane type. Such fractions are typified by a mito-chondrial preparation, readily prepared from a variety of cell types.

Reference was made in chapter 3 to the two distinct membranes which together constitute the mitochondrion. Intact mitochondria may be isolated from tissue homogenates by differential centrifugation. The

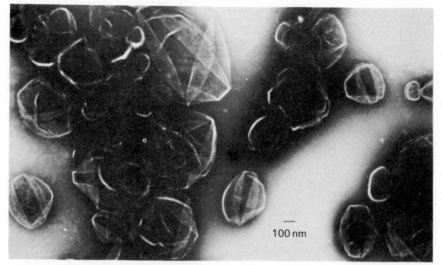

100 nm

Figure 4.5a. An electron micrograph of an osmium tetroxide-fixed negatively-stained preparation of purified mitochondrial outer membranes from rat liver. The preparation consists almost entirely of folded collapsed bags, composed of a single membrane layer. There are no sub-units or particulate material associated with the membrane. From D. F. Parsons, G. R. Williams and B. Chance, *Ann. N.Y. Acad. Sci.* (1966), **137**, 643–666. Original micrograph by courtesy of Professor Donald F. Parsons, Roswell Park Memorial Institute, Buffalo.

subsequent separation of the two mitochondrial membranes has been achieved, and their biochemical properties have been found to be quite different. Several methods have been developed for separating the membranes, mostly based on treatment of intact mitochondria with the glycoside *digitonin*. This compound interacts with the outer membrane, which may then be separated from the intact inner membrane and matrix or *mitoblast* by differential and/or density-gradient centrifugation (see figure 4.5).

Treatment of the mitoblast with a detergent allows the separation of the membrane from the soluble matrix. The outer membranes appear quite smooth, though some of the fine surface structure referred to in chapter 3 may be seen, and the inner membrane shows the characteristic knobs and stalks that are seen in intact mitochondria. Analysis of the two membranes reveals major differences in enzyme activities as shown in Table 4.1.

The inner membrane is also found to contain most of the 1,3-bis (phosphatidyl)-glycerol (cardiolipin, see chapter 5), the phospholipid that

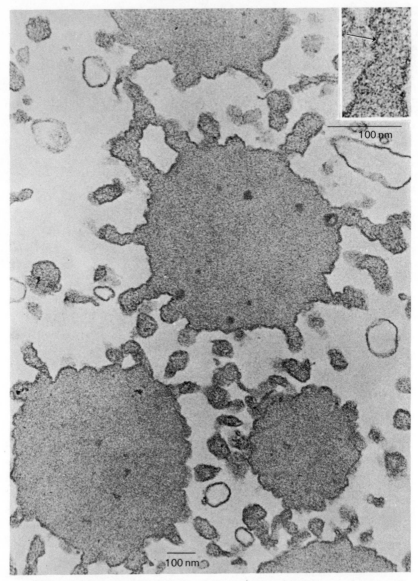

Figure 4.5b. An electron micrograph of a rat liver mitochondrial inner membrane-matrix preparation. The inner membrane appears intact, and in the inset the trilamellar nature of the membrane is seen (arrow). From C. Schnaitman and J. W. Greenawalt, *J. Cell. Biol.* (1968), **38**, 158–183. Original micrograph by courtesy of Professor John W. Greenawalt, Department of Physiological Chemistry, The Johns Hopkins School of Medicine.

Table 4.1. Enzymes of the inner and outer mitochondrial membranes

Outer membrane	Inner membrane
Monoamine oxidase	ATPase
NADH-cytochrome b_5 reductase	Carnitine acetyltransferase
Glycerophosphate acyltransferase	Steroid 11-β-hydroxylase
Hexokinase 2	Succinate dehydrogenase
Cholinephosphotransferase	Choline dehydrogenase
Acyl-Co synthetase	Cytochrome oxidase
	3-hydroxybutyrate dehydrogenase

is frequently said to be characteristic of mitochondria. Some workers have reported that the outer membrane is greatly enriched in phosphatidyl-inositol compared with the inner membrane. but others claim that this is not so. Indeed there is still considerable disagreement over the enzyme activities and composition of the two membranes, although their structural characteristics are well established.

Myelin

No account of membrane preparation would be complete without a mention of myelin. Yet myelin is in many ways the most atypical of all the membrane preparations. The characteristic multilayered appearance of the myelin sheath has been known for many years and has always attracted the attention of electron microscopists (see chapter 3). Myelin is prepared by homogenizing nervous tissue and then going through standard differential and density-gradient centrifugation procedures. Because of its very low density, a reflection of its high lipid content, myelin is often purified by flotation, i.e. if we subject a crude myelin fraction to density-gradient centrifugation, the density of the gradient can be arranged so that the myelin floats to the top, well away from any other membraneous constituents. Much valuable information on the physical properties of membranes has come from studies on purified myelin. However, myelin tells us little of membrane function. It has few characteristic enzyme activities associated with it, and appears not to have any of the very active transporting and carrier systems usually found in plasma membranes. Thus, if we wish to look at the dynamic properties of cell membranes, the study of myelin will tell us little.

Plant cell membranes

The preparation of plant cell membranes is greatly complicated by the presence of the thick polysaccharide cell wall, and so far no methods for the routine preparation of large quantities of pure plant membranes have emerged. Methods have been developed for isolating mitochondria and chloroplasts which involve an initial rather harsh disruption of the tissue, followed by differential and density-gradient centrifugation. Pure preparations of chloroplast membranes can be obtained, but in general the yield is low, considerable breakage of the organelles occurring during the initial tissue disruption.

Plasma membranes of plant cells are less well characterized than those of animal or microbial cells, but the use of *protoplasts* (see page 43) is beginning to change this. Electron microscopy of plant cell protoplasts ` treated with ferritin- or gold-labelled Concanavalin A reveals the presence of carbohydrate residues on the plasma membrane surface and we can presume that, as in the case of animal plasma membranes (see chapter 7), the residues are the oligosaccharide portions of membrane glycoproteins.

The cell membranes of micro-organisms

Micro-organisms can be produced in large quantities relatively easily and cheaply compared with animal cells. Furthermore, a homogeneous population of cell types is available, thereby circumventing many of the problems outlined above. Most micro-organisms have an extraneous coat or cell wall and, before isolation of the plasma membrane can be attempted, these outer barriers must be removed.

As described previously (page 43) this may be achieved by enzymic digestion of the wall leaving a plasma-membrane bounded *protoplast* or *spheroplast*. Pure preparations of plasma membranes can be made from protoplasts in an analogous manner to the preparation of erythrocyte ghosts. The protoplasts are put into a hypotonic medium, and the empty plasma membrane sacs collected by sedimentation. Figure 4.6 shows plasma membrane fractions prepared from bakers' yeast (*Saccharomyces cerevisiae*) by treatment with an enzyme preparation from snail gut.

The cell-envelope of the Gram-negative bacteria comprises two distinct membranes (chapter 1, page 7). Spheroplasts have been prepared from both *E. coli* and from *Salmonella typhimurium* by treatment with EDTA and lysozyme. The density of the outer membrane is higher than that of the cytoplasmic or plasma membrane, because of the presence of the polysaccharide chains of the lipopolysaccharide (page 89) that is a major constituent of the former membrane. Thus, after disruption of the spheroplasts, the two membranes can be separated by sucrose density-gradient centrifugation.

Once the problem of the cell wall has been overcome, the plasma membranes of certain micro-organisms have many attractions. The

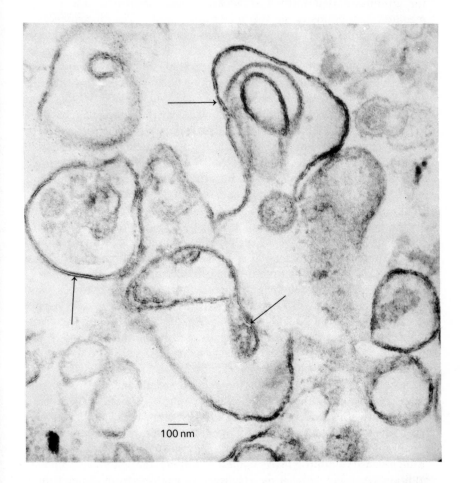

100 nm

Figure 4.6. Electron micrograph of a preparation of yeast plasma membranes purified by zonal centrifugation. The trilamellar structure of the membrane is clearly seen (arrows): from H. Suomalainen and T. Nurminen. 'Structure and Function of the Yeast Cell Envelope' in *Yeast, Mould and Plant Protoplasts, Proceedings of the Third International Symposium on Yeast Protoplasts, Salamanca*, 1972 (ed. J. R. Villanueva, I. Garcia-Acha, S. Gascon and F. Uruburu), Academic Press, London, pp. 167–185. Original micrograph by courtesy of Dr. Heikki Suomalainen, The Finnish State Alcohol Monopoly, Helsinki.

eukaryotic micro-organisms have cell membranes that show many similarities to those of plant and animal cells. One of the most convenient eukaryotic micro-organisms with which to work is yeast. Mutant strains of *Saccharomyces cerevisiae* have been isolated which have a requirement for an unsaturated fatty acid which, when supplied in the growth medium, predominates in the membrane lipids. The dependence of molecular order on fatty acid composition has been studied using spin labels (chapter 10) in the membranes of such mutants. The possibility of varying the lipid composition of a natural membrane is clearly most attractive, and considerable effort is now being directed towards this end. Strains of *Saccharomyces cerevisiae* are found which, when grown under strictly anaerobic conditions, show a growth requirement not only for an unsaturated fatty acid but also for a sterol. Protoplasts with plasma membranes enriched in oleic, linoleic and linolenic acids have been obtained, and have been found to become increasingly susceptible to osmotic lysis as the degree of unsaturation of the acyl chains, and therefore the fluidity of the membrane (see chapter 6), increases. Similarly membrane stability has been correlated with sterol composition, and the further possibility exists of simultaneously varying both sterol and fatty acid. This system presents the membrane researcher with opportunities that are not feasible with animal or higher-plant membrane preparations.

The *mycoplasmas* constitute the major order of the wall-less prokaryotes. They are bound only by a plasma membrane lacking both cell walls and mesosomes, and are accordingly sensitive to osmotic lysis, by means of which their plasma membrane can be readily isolated quite free from other membranes. The organisms are all parasites and lack many of the metabolic activities of other higher prokaryotes. Most of the mycoplasmas require long-chain fatty acids and cholesterol for growth, which means that the composition of the plasma membrane may be manipulated by changing the growth medium of the organisms. Experiments carried out with these cells provided the first clear evidence that cholesterol functions as a regulator of membrane fluidity (chapter 6). The organisms have also proved to be extremely useful in elucidating the role of high-density lipoproteins as cholesterol donors/acceptors and have in addition been widely used as a source of membranes of defined lipid composition in studies of the physical behaviour of membranes (see chapter 10).

Another micro-organism that is attractive as an experimental system is the protozoan *Tetrahymena pyriformis*. *Tetrahymena* is a typical eukaryotic cell about which we already have a considerable amount of morphological and biochemical data. Recently workers have turned to *Tetrahymena* as a model system in which to study dynamic membrane interactions. In this organism it has been clearly demonstrated that cellular metabolism may be regulated by fluidity changes in certain membranes brought about by changing the fatty acid composition of the constituent phospholipids. Thus, for example, the plasma membrane of *Tetrahymena* contains the enzyme adenylate cyclase which has a most important role in metabolic regulation in the cells of higher organisms (see chapter 9). Although its function in *Tetrahymena* is not clear, the enzyme is seen to be greatly influenced by the lipid composition of the plasma membrane, and these findings may provide further insights into the regulation of the enzyme in higher organisms. The relatively simple cellular structure of *Tetrahymena* offers the opportunity of isolation and structural characterization of the ten to fifteen different membrane types that constitute the whole cell. In this case at least, therefore, the possibility of a complete understanding of the total functions of the membranes of a simple eukaryotic cell seems not too remote.

Viral membranes

A range of RNA animal viruses, particularly the alphaviruses, the rhabdoviruses, the orthomyxoviruses and the paramyxoviruses, matures by budding from the host-cell plasma membranes, producing membrane-bound viral particles or *virions*. The membrane or *envelope* of the budding virus can be seen to be continuous with the unit membrane of the host cell in thin-section electron micrographs (see figure 4.7). Despite this apparent continuity, however, there is considerable evidence which suggests that the sections of host-cell plasma membrane that are involved in the process differ from neighbouring non-involved plasma membrane. The viral membrane has a lipid composition which is generally closely similar to that of the host-cell plasma membrane, but the protein content is quite different, and appears to be coded for by the viral genome. The envelope contains only two or three different polypeptides, and host-cell protein is not present.

Enveloped viruses formed by this process are excellent systems for the study of membrane structure and biosynthesis. Differential centrifugation in density gradients allows the preparation of pure virus particles which are free from cells and contain only protein and nucleic acid in addition to the limiting membrane. Moreover, the membrane composition can to some extent be manipulated. For instance, viral membranes with the

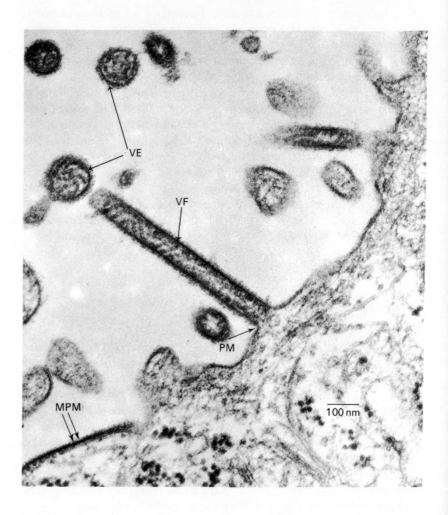

Figure 4.7. Electron micrograph of a portion of the cell surface with viral SV5 filaments, VF, in the process of budding. The circular profiles are released viral particles with a clearly defined viral membrane or envelope VE. It can be seen that the host-cell plasma membrane PM is continuous with the membrane of the viral filament. A portion of modified plasma membrane MPM is seen with surface projections and underlying nucleocapsid. From R. W. Compans and P. W. Choppin (1973). Orthomyxoviruses and paramyxoviruses in *Ultrastructure of Animal Viruses and Bacteriophages: An Atlas* (eds. A. J. Dlaton and F. Haguenau), Academic Press, New York, pp. 213–237. Original micrograph by courtesy of Dr. Richard W. Compans, The Rockefeller University, New York.

same protein but different lipids can be obtained by growing a given virus in different types of host cell, whereas membranes with similar lipids and different proteins arise from different virus types grown in the same host cell. The viral membrane proteins have been studied in some detail. It is seen that one or two types of glycoprotein are present and that, as in higher cells (see chapter 7), these project outwards from the outer face of the envelope. Again, as in higher cells, the glycoproteins are involved in recognition processes and are essential for introduction of the viral genome into the host cell; thus virions in which the extended glycopeptide portions have been removed are non-infectious.

The enveloped viruses are a unique experimental system, and many of the physical techniques described in chapter 10 have been employed using viral membranes as model systems to gain information about biological membranes in general. The role played by viral membrane protein in the initial attack on a host cell will be outlined in chapter 7.

SUMMARY

1. Before carrying out chemical analyses, pure membranes must be obtained.

2. Preparation of membranes involves rupturing the cells in such a way that the membranes are not damaged. The most common procedure employs a low shear force applied to a suspension of the tissue in iso-osmotic sucrose solution. High shear forces, enzymic digestion and gas bubble nucleation are alternative methods that have been successfully applied, particularly to plant cells and micro-organisms.

3. The membrane fractions are usually isolated by differential and density-gradient centrifugation.

4. Isolated membranes are characterized morphologically by the content of marker enzymes and by the presence of other specific constituents, such as particular lipids. Ideally a combination of analyses should be used.

5. Few membranes have been isolated in a pure form. Mammalian membranes and the plasma membranes of micro-organisms are better characterized than plant cell membranes. The erythrocyte plasma membrane, plasma membranes from liver cells, and the milk fat globule membrane represent better-characterized membrane preparations. Cultured mammalian cells provide the possibility of studying the genetic control of plasma membrane structure and function.

   The micro-organisms have an extraneous coat or cell wall that must be removed before membrane isolation can be achieved. Enzymic digestion of the coat to yield a protoplast or spheroplast has been successfully employed with the bacteria and the yeasts. The *mycoplasmas* are a particularly attractive group of micro-organisms in that they have no cell wall and the composition of their plasma membranes can be manipulated by changing the growth medium. The protozoan *Tetrahymena pyriformis* is an attractive model eukaryotic system. Its relatively simple cellular structure offers the attraction of studying some of the membrane-associated activities of higher organisms in a more amenable experimental system. The *enveloped viruses* present the opportunity of varying both the lipid and protein composition of a plasma membrane-derived fraction.

CHAPTER FIVE

MEMBRANE COMPONENTS

WHILE THE COMPOSITION OF MEMBRANES VARIES WITH THEIR SOURCE, THEY generally contain approximately 40 % of their dry weight as *lipid* and 60 % as *protein*, held together in a complex by non-covalent interactions. Usually carbohydrate is present to the extent of 1–10 % of the total dry weight. This is covalently bonded either to lipid or to protein, and the carbohydrate-containing molecules will be considered as lipids or proteins as appropriate. In addition to the above components, membranes contain some 20 % of their total weight as water, which is tightly bound and essential to the maintenance of their structure.

Lipids

Lipids are water-insoluble organic substances which can, in general, be extracted by non-polar solvents such as chloroform, ether and benzene, and analysed by thin-layer chromatography on silicic acid.

Membrane lipids are polar lipids which are also referred to as *amphipathic*, meaning that they incorporate both a hydrophobic tail and a hydrophilic head group within the molecule. The hydrophobic and hydrophilic regions can be bridged by a *glycerol* moiety, by a *sphinganine derivative* or homologue or, finally, within a *sterol* molecule.

In the glycerol derivatives, the hydrophobic tail commonly consists of two long-chain fatty acids esterified to two hydroxy-groups of glycerol, while the third, primary, hydroxy-group carries a hydrophilic grouping. Such a diacylglycerol derivative is shown in figure 5.1. The hydrophobic tail can also contain long-chain fatty aldehyde and alcohol molecules, attached to the glycerol hydroxy-groups by other linkages, as shown in the legend to figure 5.1.

Sphinganine derivatives and homologues (*sphingoids*) also contain a terminal hydroxy-group which can carry a hydrophilic grouping. They have additionally an amino group and a long aliphatic chain built into

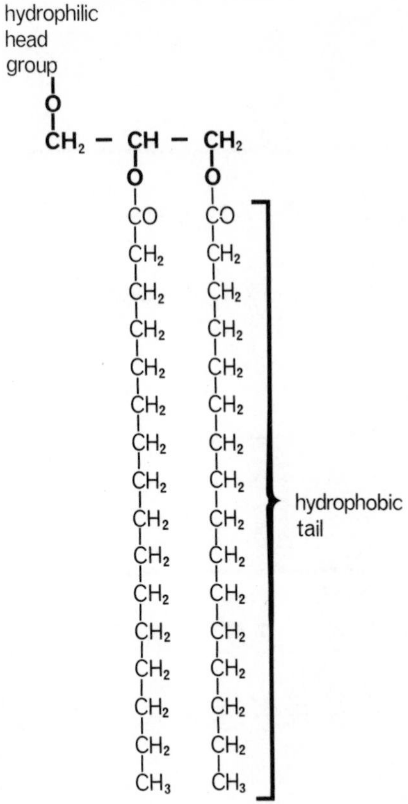

hydrophilic
head
group
|
O
|
CH₂ — CH — CH₂
| |
O   O
| |
CO CO
| |
CH₂ CH₂
| |
CH₂ CH₂
| |
CH₂ CH₂
| |
CH₂ CH₂
| |
CH₂ CH₂
| |
CH₂ CH₂
| |
CH₂ CH₂   hydrophobic
| | tail
CH₂ CH₂
| |
CH₂ CH₂
| |
CH₂ CH₂
| |
CH₂ CH₂
| |
CH₂ CH₂
| |
CH₂ CH₂
| |
CH₃ CH₃

Figure 5.1. Polar lipids based on a glycerol backbone (shown in bold type). The hydrophobic tail is shown containing two C_{16} saturated fatty acids, but chain lengths may vary and unsaturated acids also occur. Moreover, long-chain fatty aldehydes and alcohols may also be linked to glycerol [by O-(1-alkenyl) (—O—CH=CH—) and O-alkyl (—O—CH₂—CH₂—) linkages respectively]. The preferred conformation of the polymethylene chains is that of a planar zig-zag (figure 5.4).

the basic structure, which together with a long-chain fatty acid (attached to the amino group via an amide linkage) constitute the hydrophobic tail of the molecule. Generalized structures of polar lipids based on sphinganine, *trans*-4-sphingenine (sphingosine) and 4-D-hydroxysphinganine (phytosphingosine), the three major sphingoids, are shown in figure 5.2. Such polar lipids are generally classed as *sphingolipids*.

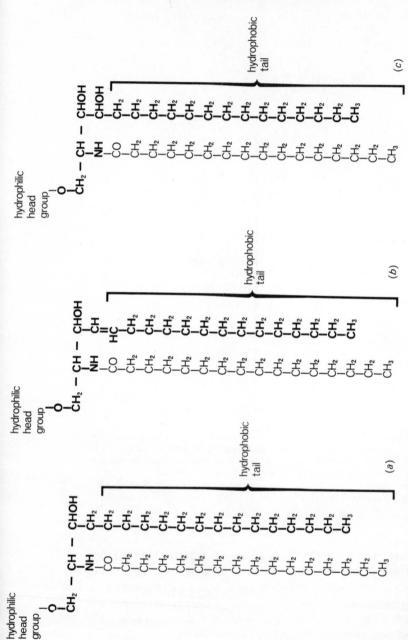

Figure 5.2. Polar lipids based on sphinganine derivatives or sphingoids (shown in bold type).
(a) sphinganine;
(b) *trans*-4-sphingenine (sphingosine);
(c) 4-D-hydroxysphinganine (phytosphingosine).
The fatty acids may be of various chain lengths and may be unsaturated. In addition, homologues of the sphinganine derivatives containing chains of different lengths and degrees of unsaturation occur.

The hydrophobic tails of polar lipids

Most models proposed for membrane structures (chapter 6) incorporate a lipid bilayer arrangement, such as that shown schematically in figure 5.3. There the long aliphatic chains are sequestered within the interior of

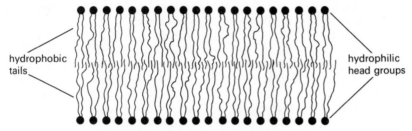

Figure 5.3. Bilayer arrangement of polar lipids.

the membrane, and all the hydrophilic head groups face the aqueous environments. A hydrophobic or apolar group in water causes extensive rearrangement of water molecules in its vicinity, leading to a thermo-dynamically unfavourable loss of entropy. Accordingly, apolar groups tend to cluster together, excluding water, and so reduce the free energy of the system. The major driving force for this hydrophobic association results, therefore, from entropic factors rather than from specific apolar-apolar interactions.

The hydrophobic tails of polar lipids are made up of the long aliphatic chains of fatty acids, aldehydes or alcohols, of sphinganine derivatives, or of sterol molecules. The composition of the hydrophobic region of a bilayer of polar lipid is accordingly capable of considerable variation, and it is probable that the physical properties of the bilayer as a whole can be subtly controlled by structural variations in the hydrophobic tails. An indication of the different structural possibilities and of their effects on the membrane can be obtained from consideration of the long-chain fatty acids, which are the commonest components of the hydrophobic tails of polar lipids. Fatty acids of higher plants and animals contain 14–24 carbon atoms, with chains of 16 and 18 predominating, and can be saturated or unsaturated. Monounsaturated acids have a double bond between C-9 and C-10. Further double bonds, if present, usually occur in the chain between carbon 10 and the terminal methyl group in such a way that two double bonds are separated by a methylene group (i.e. $CH{=}CH-CH_2-CH{=}CH$) and never conjugated (as in $CH{=}CH-CH{=}CH-$). Plant fatty acids in general are highly unsaturated and

Table 5.1. Fatty acids commonly occurring in higher plants and animals.

Numerical symbol*	Structure $H_3C-[R]_n-COOH$	Systematic name	Trivial name
14:0	$-[CH_2]_{12}-$	tetradecanoic	myristic
16:0	$-[CH_2]_{14}-$	hexadecanoic	palmitic
16:1(9)	$-[CH_2]_5CH=CH[CH_2]_7-$	9-hexadecenoic	palmitoleic
18:0	$-[CH_2]_{16}-$	octadecanoic	stearic
18:1(9)	$-[CH_2]_7CH=CH[CH_2]_7-$	cis-9-octadecenoic	oleic
18:2(9,12)	$-[CH_2]_3[CH_2CH=CH]_2[CH_2]_7-$	cis,cis-9,12-octadecadienoic	linoleic
18:3(9,12,15)	$-[CH_2CH=CH]_3[CH_2]_7-$	9,12,15-octadecatrienoic	(9,12,15)-linolenic
20:0	$-[CH_2]_{18}-$	icosanoic	arachidic
20:4(5,8,11,14)	$-[CH_2]_3[CH_2CH=CH]_4[CH_2]_3-$	5,8,11,14-icosatetraenoic	arachidonic
22:0	$-[CH_2]_{20}-$	docosanoic	behenic
24:0	$-[CH_2]_{22}-$	tetracosanoic	lignoceric
24:1(15)	$-[CH_2]_7CH=CH[CH_2]_{13}-$	cis-15-tetracosenoic	nervonic

* A:B(C,D) where A = number of carbon atoms; B = number of double bonds; C,D = positions of double bonds.

9,12,15-octadecatrienoic acid (linolenic acid) is often a major component. Commonly occurring fatty acids of higher organisms are shown in Table 5.1. Bacterial fatty acids are generally 10–20 carbon atoms long and can also be saturated or unsaturated. The unsaturated acids contain only a single double bond, usually between C-11 and C-12, and occur predominantly in Gram-negative bacteria often together with cyclo-

Table 5.2. Fatty-acid types commonly occurring in bacteria.

$CH_3(CH_2)_nCOOH$	saturated fatty acids
$\underset{\displaystyle CH_3CH(CH_2)_nCOOH}{\overset{\displaystyle CH_3}{\vert}}$	branched chain *iso* fatty acids
$\underset{\displaystyle CH_3CH_2CH(CH_2)_nCOOH}{\overset{\displaystyle CH_3}{\vert}}$	branched chain *anteiso* fatty acids
$CH_3(CH_2)_5C\underset{\displaystyle H}{\overset{\displaystyle \diagup CH_2 \diagdown}{\vert}}\!\!\!\!\!\!\!\!\!\!\!\!\!\!C(CH_2)_9COOH$	cyclopropane fatty acid (lactobacillic)
$CH_3(CH_2)_5CH=CH(CH_2)_9COOH$	monounsaturated acid (*cis*-vaccenic acid)

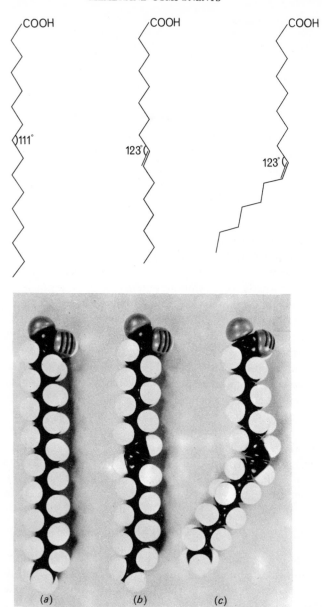

Figure 5.4. Minimum energy conformations of (*a*) saturated fatty acid, (*b*) fatty acid containing a *trans* double bond, (*c*) fatty acid containing a *cis* double bond.

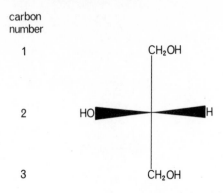

Figure 5.5. Stereospecific numbering of glycerol.

propane fatty acids (Table 5.2). Branched-chain fatty acids (Table 5.2)
are widely found in Gram-positive bacteria.

Although free rotation is possible about each carbon-carbon bond in
saturated fatty acids, the minimum-energy conformation is that of the
planar zig-zag shown in figure 5.4a. Such extended chains might be
expected to fit well into an ordered crystalline type of array within a
bilayer structure. The presence of a *trans* double bond in the chain makes
little difference to the overall conformation (figure 5.4b). Naturally-
occurring unsaturated fatty acids almost always contain *cis* double
bonds, however, which impart an overall bend of approximately 30° to
the chain (figure 5.4c). Incorporation of such kinked molecules into the
ordered array of a bilayer will disrupt the array and tend to make the
membrane more fluid. This effect is reflected in the melting-points of
naturally-occurring unsaturated fatty acids which, on average, melt
50 °C lower than those of the corresponding saturated acids (see also
chapter 10). Similar effects on membrane fluidity may be brought about
by the branched-chain and cyclopropane fatty acids of bacteria.

Phospholipids

Phospholipid is a term applied to any lipid containing phosphoric acid as
a mono- or di-ester, and the commonest examples are based on the
generalized polar lipids shown in figures 5.1 and 5.2 in which the
hydrophilic head group is attached via a phosphate di-ester linkage.

Glycerphospholipids are phospholipids in which a phosphate ester
of glycerol contains at least one *O*-acyl, *O*-alkyl or *O*-(1-alkenyl) group

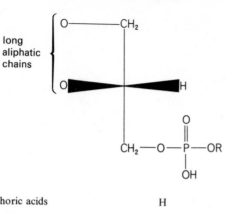

sn-Glycero-3-phosphoric acids   H

sn-Glycero-3-phosphocholines $-CH_2CH_2\overset{+}{N}(CH_3)_3$

sn-Glycero-3-phosphoethanolamines $-CH_2CH_2NH_2$

sn-Glycero-3-phosphoserines $-CH_2CH(NH_2)COOH$

sn-Glycero-3-phosphoinositols

sn-Glycero-3-phosphoglycerols $-CH_2CHOH-CH_2OH$

1,3-bis (sn-glycero-
  3-phospho)-glycerols

$$-CH_2-CHOH-CH_2-O-\overset{\overset{\textstyle O}{\|}}{\underset{\underset{\textstyle OH}{|}}{P}}-O-CH_2-CH-CH_2$$

long
aliphatic
chains

sn-Glycero-3-phosphoaminoacylglycerols $-CH_2-CHOH-CH_2-OR^*$

Figure 5.6. The major glycerophospholipids are based on the general structure shown.
R*=L-alanyl, L-lysyl, L-ornithyl or L-arginyl.

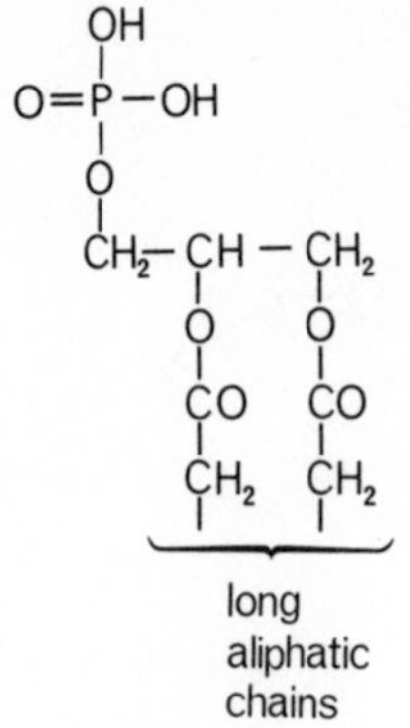

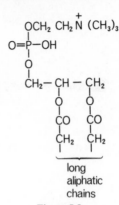

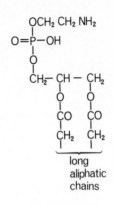

Figure 5.7.
Phosphatidic acid
(1,2-diacyl-*sn*-glycero-
3-phosphoric acid).

Figure 5.8.
Phosphatidylcholine
(1,2-diacyl-*sn*-glycero-
3-phosphocholine).

Figure 5.9.
Phosphatidylethanolamine
(1,2-diacyl-*sn*-glycero-3-
phosphoethanolamine).

attached to the glycerol residue. The phosphate ester grouping is usually attached to one of the primary hydroxyl groups of the glycerol moiety, i.e. linking the hydrophilic head group of figure 5.1. Which of the two stereochemically non-identical primary hydroxyl groups is so involved is best designated by the *stereospecific numbering* system. In this system, when glycerol is drawn in the Fischer projection with the C-2 hydroxyl group on the left, the carbon atoms are numbered as shown in figure 5.5. When the stereospecific numbering system is being applied, the prefix *sn* is used. Glycerophospholipids usually contain an *sn*-glycerol 3-phosphate moiety either as a monoester (figure 5.6, R=H) or as a diester linked to one of the remaining R groups listed in figure 5.6.

The diacyl derivative of *sn*-glycerol 3-phosphoric acid (figure 5.6, R=H; figure 5.7) is named (3-*sn*)-phosphatidic acid, commonly abbreviated to phosphatidic acid, and plays a key role in the biosynthesis of most glycerophospholipids. It is accordingly widely distributed, occurring at a low level (1–5 % of total phospholipids) in many tissues.

The diacyl derivatives of the major glycerophospholipids are named as derivatives of (3-*sn*)-phosphatidic acid. Thus the most abundant glycerophospholipids in eukaryotic micro-organisms are phosphatidylcholine (1,2-diacyl-*sn*-glycero-3-phosphocholine) (figure 5.8) and phosphatidylethanolamine (1,2-diacyl-*sn*-glycero-3-phosphoethanolamine) (figure 5.9). It has been recommended that the old trivial names for these compounds, lecithin and cephalin respectively, should be abandoned. Phosphatidylcholine is usually the major phospholipid in mammalian, plant and

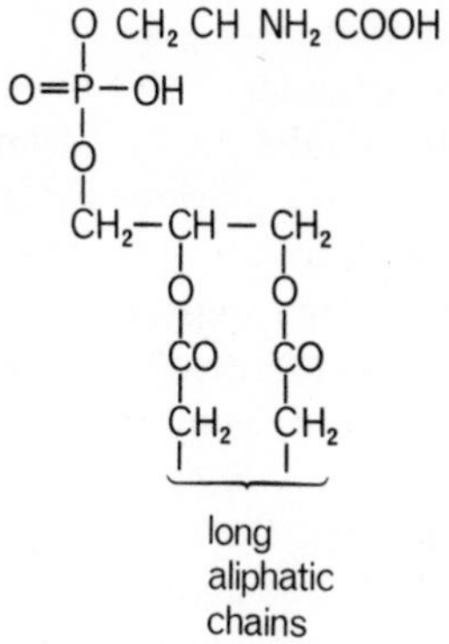

Figure 5.10. Phosphatidylserine (1,2-diacyl-*sn*-glycero-3-phosphoserine).

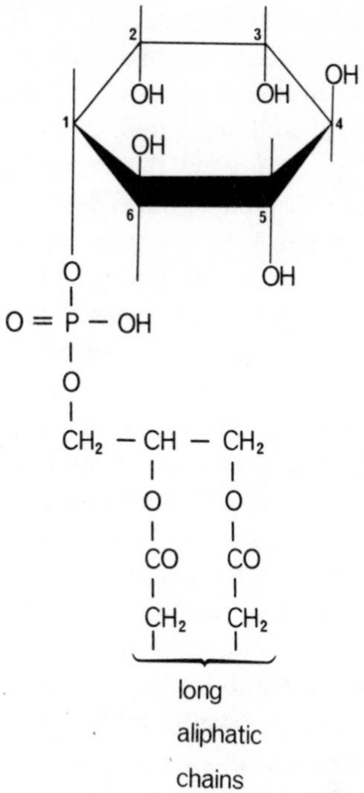

Figure 5.11. Phosphatidylinositol (1,2-diacyl-*sn*-glycero-3-phosphoinositol).

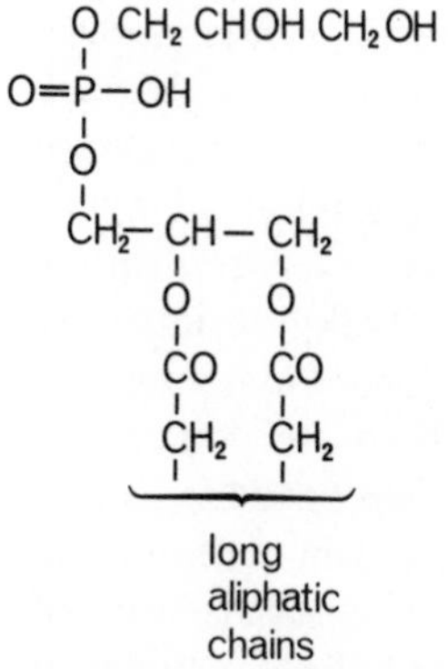

Figure 5.12. Phosphatidylglycerol (1,2-diacyl-*sn*-glycero-3-phosphoglycerol).

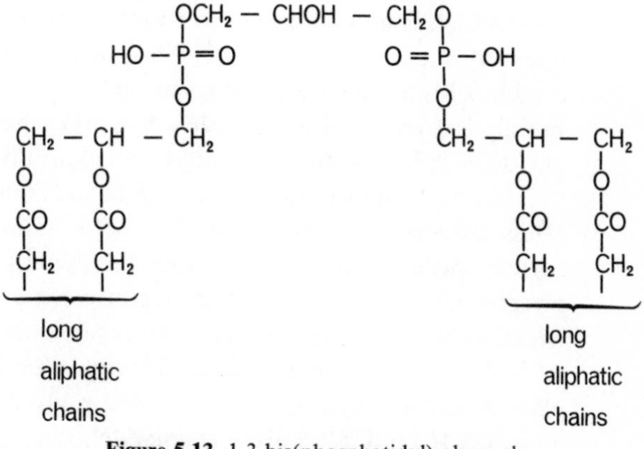

Figure 5.13. 1,3-bis(phosphatidyl)-glycerol, [1,3-bis(1,2-diacyl-*sn*-glycero-3-phospho) glycerol], (cardiolipin).

fungal membranes, but is comparatively rare in bacteria in which phosphatidylethanolamine is often a major component (see page 88). In these diacyl derivatives fatty-acid substitution falls into a pattern with unsaturated acids mainly substituted at C-2 of the glycerol moiety.

The monoacyl derivative lysophosphatidylcholine, and to a lesser extent lysophosphatidylethanolamine, have been reported as minor components in many membranes. There is some evidence that these glycerophospholipids, which can actually disrupt membrane structure, arise from phospholipase action.

The diacyl derivatives phosphatidylserine (figure 5.10) and phosphatidylinositol (figure 5.11) are widely distributed as minor (usually less than 10%) phospholipid components of eukaryotic membranes. Phosphatidylinositol occurs as its 4-phosphate and 4,5-diphosphates (figure 5.11), highest concentrations of which have been reported in nervous tissue (largely myelin). Phosphatidylserine, a biosynthetic precursor of phosphatidylethanolamine, is present, but does not normally accumulate, in bacteria, whereas phosphatidylinositol is virtually absent from prokaryotic membranes.

Phosphatidylglycerol (figure 5.12) is probably the most widely-occurring phospholipid in bacteria, being particularly abundant in the membranes of Gram-positive organisms. It is also a major phospholipid constituent of chloroplast membranes, although it is present as only a minor component of plant and animal (mainly mitochondrial) membranes generally and is rare in fungi. Phosphatidylglycerol is a biosynthetic precursor of 1,3-bis(phosphatidyl)-glycerol (cardiolipin, figure 5.13) which is accordingly also found largely in bacterial, mitochondrial and chloroplast membranes. Phosphatidyl 3-O-aminoacyl-sn-glycerols (figure 5.14) are also derived from phosphatidylglycerol. They have been mainly located in Gram-positive bacteria where the L-lysyl and L-alanyl and to a lesser extent L-ornithyl and L-arginyl esters have been found.

Glycerophospholipids containing O-(1-alkenyl) or O-alkyl residues are often detected as minor membrane components accompanying the corresponding diacyl derivatives. Glycerophospholipids in which the glycerol moiety bears a 1-alkenyl group are known as *plasmalogens* and when, as is usual, the O-(1-alkenyl) residue occurs at C-1 in combination with an O-acyl group at C-2, the plasmalogen is named as a derivative of plasmenic acid (figure 5.15a). Thus the glycerophospholipids shown in figures 5.16 and 5.17 are plasmenylcholine (formerly choline plasmalogen) and plasmenylethanolamine (formerly ethanolamine plasmalogen) respectively. These are the most widely-reported plasmalogens and have

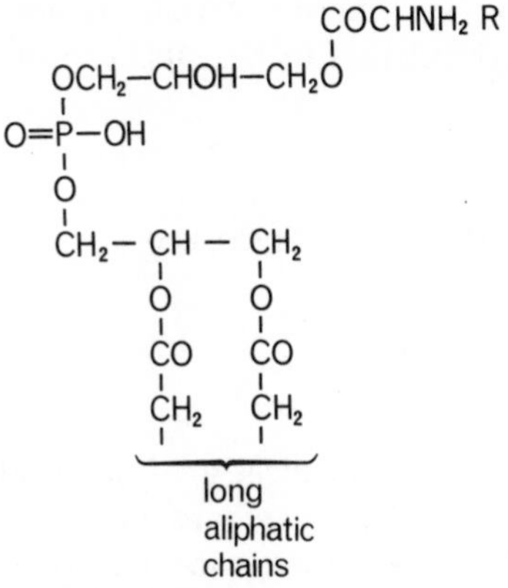

COCHNH₂ R
|
OCH₂—CHOH—CH₂O
|
O=P—OH
|
O
|
CH₂— CH — CH₂
| |
O   O
| |
CO   CO
| |
CH₂   CH₂

long
aliphatic
chains

Figure 5.14. Phosphatidyl 3-O-aminoacyl-sn-glycerol [1,2-diacyl-sn-glycero-3-phospho(3′-aminoacyl-sn-glycerol)].

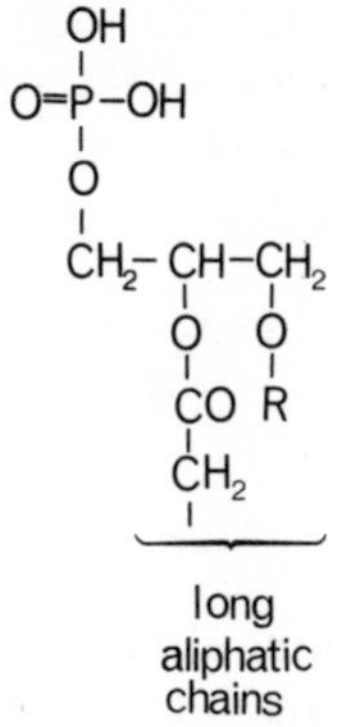

OH
|
O=P—OH
|
O
|
CH₂— CH—CH₂
| |
O   O
| |
CO   R
|
CH₂

long
aliphatic
chains

Figure 5.15. (a) R=—CH=CH—, plasmenic acid; (b) R=—CH₂—CH₂—, plasmanic acid.

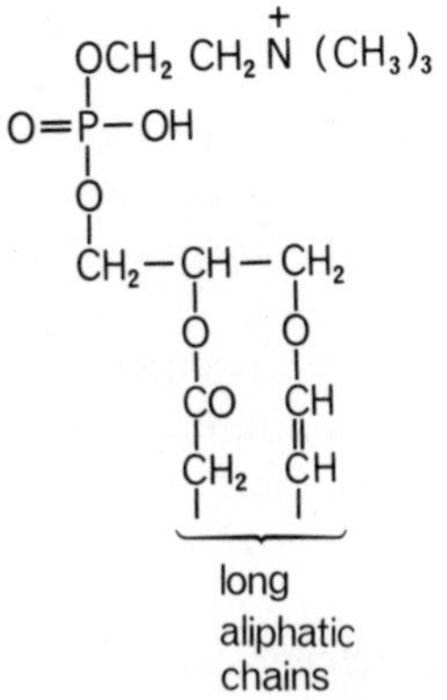

OCH₂ CH₂ N⁺(CH₃)₃
|
O=P—OH
|
O
|
CH₂—CH—CH₂
| |
O   O
| |
CO   CH
| ‖
CH₂ CH

long
aliphatic
chains

Figure 5.16. Plasmenylcholine (choline plasmalogen).

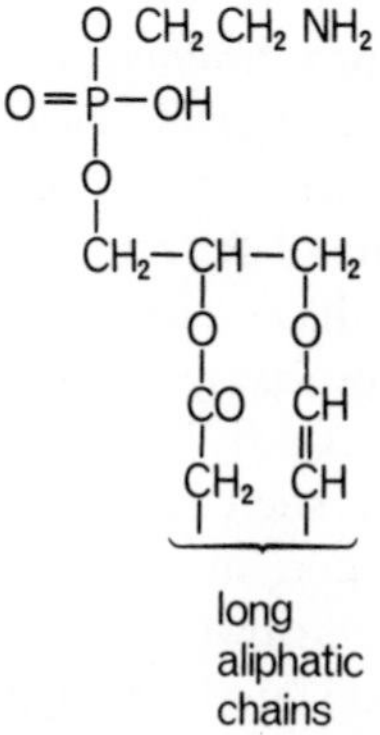

O CH₂ CH₂ NH₂
|
O=P—OH
|
O
|
CH₂—CH—CH₂
| |
O   O
| |
CO   CH
| ‖
CH₂ CH

long
aliphatic
chains

Figure 5.17. Plasmenylethanolamine (ethanolamine plasmalogen).

been detected in relatively high concentrations in heart muscle and in myelin. Bacteria (mainly anaerobes) also show a variable plasmalogen content.

2-O-Acyl-1-O-alkyl-sn-glycero-3-phospholipids [plasmanyl derivatives, named after plasmanic acid (figure 5.15b)] and the 1,2-di-O-alkyl analogues have also been characterized from a number of sources. Extremely halophilic bacteria contain glycerophospholipids that are exceptional in that long-chain alcohols such as dihydrophytyl alcohol (figure 5.18) are substituted via *O*-alkyl linkages to C-2 and C-3 of the *sn*-glycerol moiety.

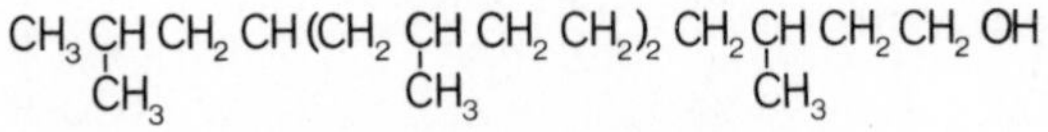

$$CH_3 \underset{\underset{CH_3}{|}}{CH} CH_2\, CH(CH_2\, \underset{\underset{CH_3}{|}}{CH}\, CH_2\, CH_2)_2\, CH_2\, \underset{\underset{CH_3}{|}}{CH}\, CH_2\, CH_2\, OH$$

Figure 5.18. Dihydrophytyl alcohol.

The stability of ether linkages hinders the chemical analysis of alkyl glycerophospholipids which, like the plasmalogens, are not easy to separate from the more abundant diacyl derivatives. In view of these difficulties it is possible that both the *O*-alkyl and *O*-(1-alkenyl) glycerophospholipids are more common than has been reported.

The second major class of phospholipids is that of the *phosphosphingolipids*, which are based on the general polar lipid structures shown in figure 5.2. where the hydrophilic head group is a phosphate diester. With recent improvements in separation techniques, the number of known sphinganine derivatives (sphingoids) has greatly increased. The three main structural types are shown in figure 5.2. Sphingosine (*trans*-4-sphingenine) is the major sphingoid of animal phosphosphingolipids, which also have a low content of sphinganine, whereas that of plants and fungi is probably 4-D-hydroxysphinganine (phytosphingosine). Until relatively recently bacteria were thought not to contain phosphosphingolipids, but such lipids containing sphinganine and some of its less common derivatives have now been characterized in *Bacteroides melaninogenicus*.

The *N*-acylated sphinganine derivative is referred to as a ceramide (figure 5.19). Sphingomyelin, ceramide 1-phosphocholine (figure 5.20), is the best-known phosphosphingolipid, occurring primarily in mammalian cells where it is generally present in highest concentrations in plasma membranes. The fatty acids of sphingomyelin are largely saturated and are often longer [e.g. tetracosanoic (lignoceric, 24:0) and *cis*-15-tetracosenoic (nervonic, 24:1) acids in myelin] than in the glycerophospholipids. Ceramide 1-phosphoinositol derivatives containing complex oligosaccharides linked to the inositol moiety occur in plants and fungi, while some bacteria have now been shown to contain phosphosphingolipids carrying 'bacterial-type' head groups (e.g. ethanolamine and glycerol, but not choline).

Space-filling models of a range of phospholipids are shown in figure 5.21.

The phospholipids are the most polar of the lipids. They all carry a phosphate acidic grouping with pK_a 1–2 which has a negative charge at pH 7·0. Choline and ethanolamine additionally have amino groups with pK_a values of 13 and 10 respectively, and are positively charged at neutral pH, giving rise to dipolar zwitterions with no net charge. Serine has one negative and one positive charge at physiological pH (7·0), and phospholipids containing serine consequently normally carry a single net negative charge, as do those containing inositol and glycerol. There is, therefore, a considerable range of hydrophilic head groups available among the phospholipids, varying in size, shape, polarity and charge. Just as variations in chain length and degree of unsaturation of the aliphatic chains of polar lipids might be expected to influence the

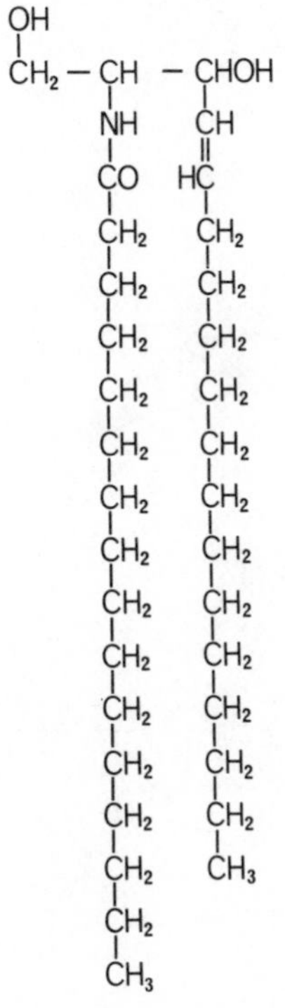

Figure 5.19. A ceramide based on *trans*-4-sphingenine (sphingosine).

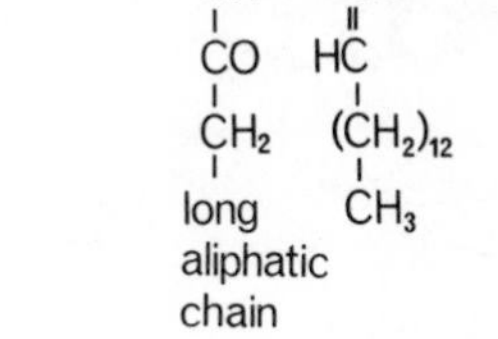

Figure 5.20. Ceramide 1-phosphocholine (sphingomyelin).

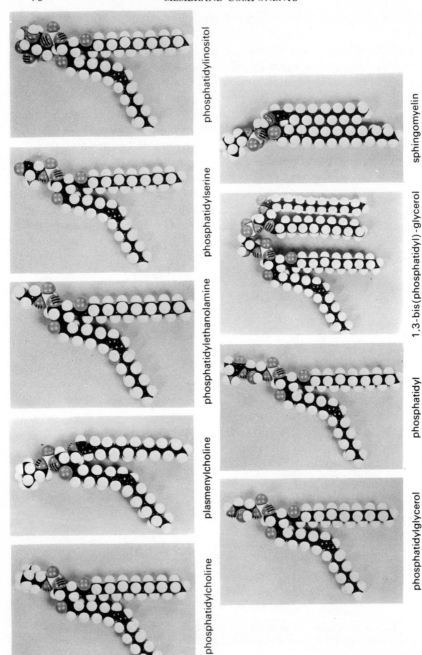

Figure 5.21. Space-filling models of phospholipids.

phosphatidylinositol

phosphatidylserine

phosphatidylethanolamine

plasmenylcholine

phosphatidylcholine

sphingomyelin

1,3-bis(phosphatidyl)-glycerol (cardiolipin)

phosphatidyl 3-aminoacyl-*sn*-glycerol

phosphatidylglycerol

properties of the lipid bilayers, so might the overall properties of the bilayer differ according to the proportions of the various hydrophilic head groups.

Glycolipids

The range of hydrophilic head groups of polar lipids is further extended by the class of *glycolipids* in which the head group is attached via the glycosidic linkage of a sugar molecule rather than by a phosphate ester bond, as in the phospholipids.

Glycoglycerolipids are glycolipids containing one or more glycerol residues, and the commonest are those (formerly called glycosyl diglycerides) based on the general polar lipid structure of figure 5.1. The monogalactosyl derivative, 1,2-diacyl-3-β-D-galactosyl-*sn*-glycerol (formerly monogalactosyl diglyceride) (figure 5.22) is a major lipid of chloroplast membranes in which the digalactosyl (figure 5.23) and sulphoquinovosyl (figure 5.24) derivatives are also found.

The disaccharide derivatives are widespread in Gram-positive bacterial membranes. The most widespread of such glycoglycerolipids contain α-glucosyl-α-glucosyl, β-glucosyl-β-glucosyl, α-galactosyl-α-glucosyl, β-galactosyl-β-galactosyl and α-mannosyl-α-mannosyl units. Glycoglycerolipids are not generally found among the lipids of Gram-negative bacteria.

Although the monogalactosyl compound (figure 5.22) has been reported in mammalian brain, glycolipids of animal membranes are largely polar lipids of the general types shown in figure 5.2. In these *glycosphingolipids*, the hydrophilic head group is attached to the terminal hydroxymethyl grouping of a ceramide by a glycosidic linkage, and the resulting compounds are referred to as glycosylceramides. Such compounds occur in brain, spleen, erythrocytes, kidney and liver, where they are located in the plasma membranes. The glycosyl moiety can vary from a single galactose or glucose unit to complex oligosaccharides containing D-galactose, D-glucose, D-galactosamine, D-glucosamine, L-fucose (6-deoxy-L-galactose) and sialic acid (*N*-acetylneuraminic acid or *N*-glycolylneuraminic acid).

1-β-D-Galactosylceramide (figure 5.25) and its glucosyl analogue are both commonly referred to as *cerebrosides*. Galactosylceramide, together with its C-3 (of galactose) sulphate ester, constitutes a significant component of brain lipid, especially in myelin. Glucosylceramide, on the other hand, is predominant in non-neuronal tissue, being widely distributed in

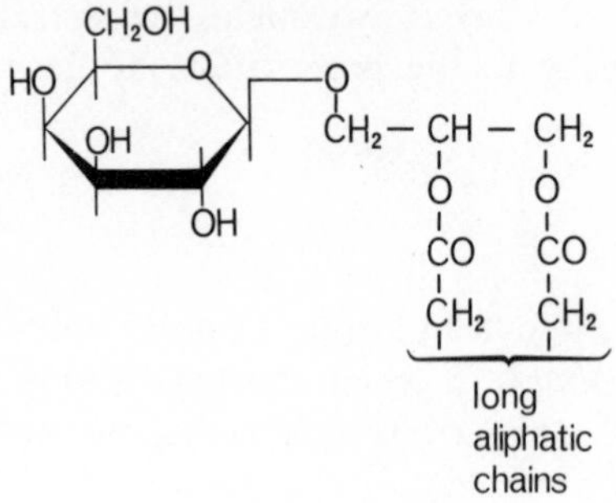

Figure 5.22. 1,2-Diacyl-3-β-D-galactosyl-*sn*-glycerol (formerly monogalactosyl diglyceride).

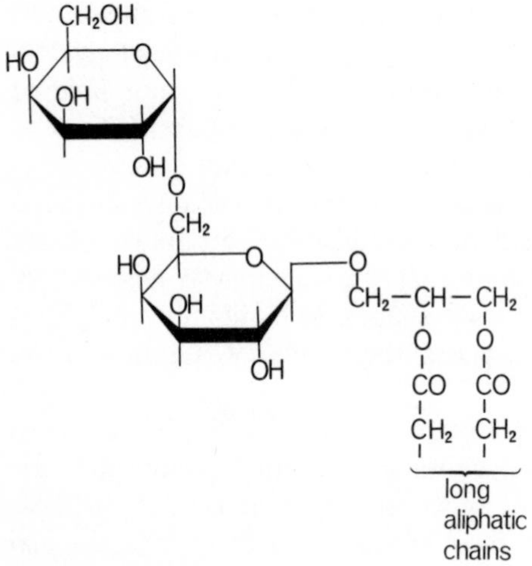

Figure 5.23. 1,2-Diacyl-3-[α-D-galactosyl (1–6) β-D-galactosyl)]-*sn*-glycerol.

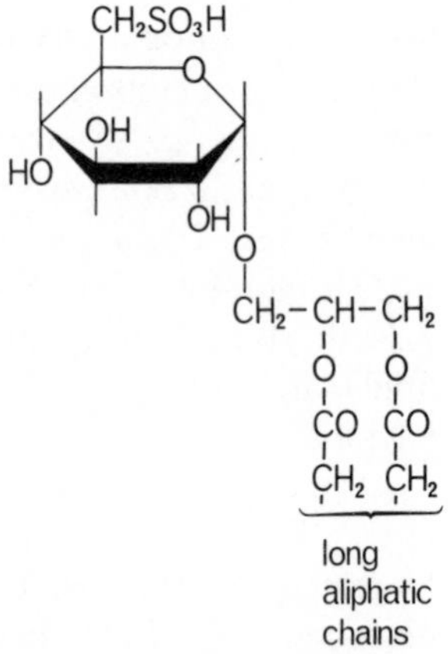

Figure 5.24. 1,2-Diacyl-3-(6-sulpho-α-D-quinovosyl)-*sn*-glycerol (N.B. quinovose is 6-deoxyglucose).

the plasma membranes of most mammalian cells. Cerebrosides often contain longer-chain fatty acids, the most abundant of which are tetracosanoic (lignoceric, 24:0), 2-hydroxytetracosanoic (cerebronic or α-hydroxylignoceric) and *cis*-15-tetracosenoic (nervonic, 24:1) acids.

The structures and nomenclature of some more complex glycosphingolipids are shown in Table 5.3. Most of the structures listed in Table 5.3 are derived from lactosylceramide (figure 5.26) by addition of further sugar units such as α-galactosyl (in the globo- and isoglobo- series),

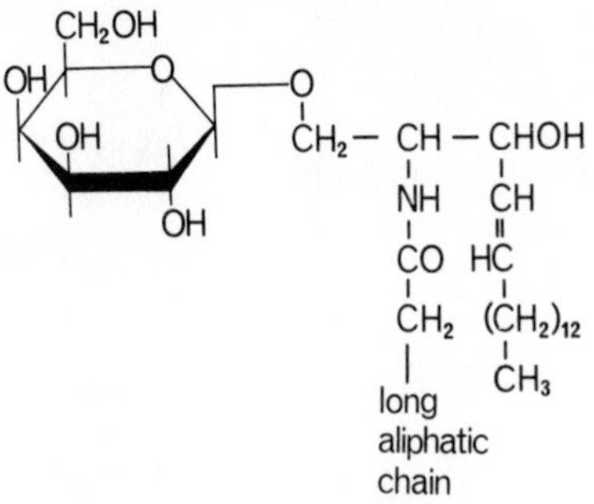

Figure 5.25. β-D-Galactosylceramide.

Table 5.3. Structures of some glycosphingolipids and trivial names of their oligosaccharides. Cer = ceramide; Glc = D-glucose; Gal = D-galactose; GlcNAc = N-acetyl-D-glucosamine; GalNAc = N-acetyl-D-galactosamine. Linkages Glc Cer and Gal Cer are in all cases $\beta1$–1.

Glycolipid structure	Trivial name of Oligosaccharide*
Gal(α1–4)Gal(β1–4)GlcCer	Globotriaose
GalNAc(β1–3)Gal(α1–4)Gal(β1–4)GlcCer	Globotetraose
Gal(α1–3)Gal(α1–4)GlcCer	Isoglobotriaose
GalNAc(β1–3)Gal(α1–3)Gal(β1–4)GlcCer	Isoglobotetraose
Gal(β1–4)Gal(β1–4)GlcCer	Mucotriaose
Gal(β1–3)Gal(β1–4)Gal(β1–4)GlcCer	Mucotetraose
GlcNAc(β1–3)Gal(β1–4)GlcCer	Lactotriaose
Gal(β1–3)GlcNAc(β1–3)Gal(β1–4)GlcCer	Lactotetraose
Gal(β1–4)GlcNAc(β1–3)Gal(β1–4)GlcCer	Neolactotetraose
GalNAc(β1–4)Gal(β1–4)GlcCer	Gangliotriaose
Gal(β1–3)GalNAc(β1–4)Gal(β1–4)GlcCer	Gangliotetraose
Gal(α1–4)GalCer	Galabiose
Gal(1–4)Gal(α1–4)GalCer	Galatriaose
GalNAc(1–3)Gal(1–4)Gal(α1–4)GalCer	N-Acetylgalactosaminylgalatriaose

*The name of the glycolipid is formed by converting -ose to -osyl, followed by ceramide, without space, e.g. globotriaosylceramide: from 'The Nomenclature of Lipids,' IUPAC-IUB Commission on Biochemical Nomenclature (1978), *Biochem. J.*, **171**, 21–35.

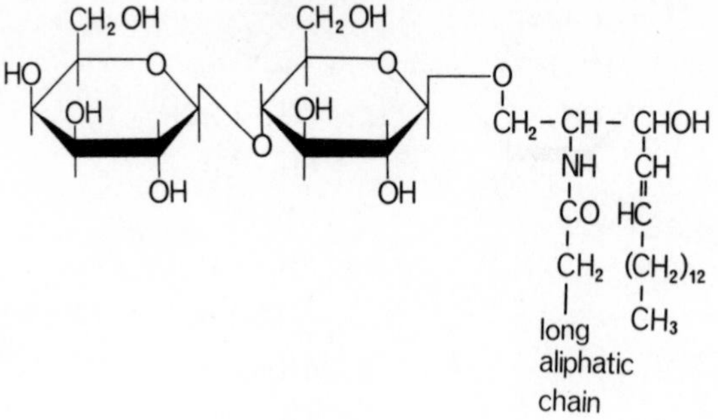

Figure 5.26. β-D-Galactosyl (1–4) -D-glucosylceramide (lactosylceramide).

Table 5.4. Structures of the major gangliosides of mammalian tissues. NANA $= N$-acetylneuraminic acid and other symbols are as in Table 5.3. Nomenclature is that of Svennerholm, *J. Neurochem.* (1963) **10**, 613–623 in which G = ganglioside, M = monosialo, D = disialo, T = trisialo, and arabic numerals indicate sequence of migration in thin-layer chromatograms.

Ganglioside structure	Designation
NANA(α2–3)Gal(β1–4)GlcCer	G_{M3}
GalNAc(β1–4)Gal(β1–4)GlcCer 3 \| NANAα2	G_{M2}
Gal(β1–3)GalNAc(β1–4)Gal(β1–4)GlcCer 3 \| NANAα2	G_{M1}
Gal(β1–3)GalNAc(β1–4)Gal(β1–4)GlcCer 3 3 \| \| NANAα2 NANAα2	G_{D1a}
Gal(β1–3)GalNAc(β1–4)Gal(β1–4)GlcCer 3 \| NANA(α2–8)NANAα2	G_{D1b}
Gal(β1–3)GalNAc(β1–4)Gal(β1–4)GlcCer 3 3 \| \| NANAα2 NANA(α2–8)NANAα2	G_{T1b}

β-galactosyl (muco-series), N-acetyl-β-glucosaminyl (lacto- and neolacto-series) or N-acetyl-β-galactosaminyl (ganglio-series). Globoside I (globotetraosylceramide) is the major glycolipid of human erythrocyte membrane where its lineas 1-3 linked N-acetyl-α-galactosaminyl derivative (Forssman glycolipid) also occurs, while the ABH blood group determinants of the red cell are all complex glycosphingolipids having a neolactotetraosylceramide core (chapter 7). Blood group-active glycosphingolipids containing as many as 40 monosaccharide residues per molecule have recently been characterized from human erythrocyte membrane.

Gangliosides are glycosphingolipids carrying one or more sialic acid residues and are present in the plasma membranes of many mammalian cells, especially nerve cells. The structures of the major gangliosides of mammalian tissues (Table 5.4) can be seen to be derived from lactosylceramide (e.g. G_{M3}), gangliotriosylceramide (e.g. G_{M2}) or gangliotetraosylceramide (e.g. G_{M1}, G_{D1a}, G_{D1b}, G_{T1b}) by addition of sialyl residues. The structure of G_{D1a} is shown in detail in figure 5.27.

In higher plants and fungi, glycosphingolipids of the above types are represented only by the simple cerebrosides.

A further minor class of glycolipids is that of the *steryl glycosides* in which a monosaccharide is glycosidically linked to the hyroxyl group of a sterol (see next section). These polar lipids are found as minor components in animal and plant membranes.

Space-filling models of glycolipids are shown in figure 5.28. It is likely that the sugar residues are exposed on the outer surface of the plasma membranes in which the glycolipids are mostly found. Particularly the more complex oligosaccharide residues of the glycosphingolipids might be expected to contribute to the specific nature of cell surfaces. Blood group specificities, for instance, are known to be associated with glycolipids in the erythrocyte membrane. Cell surface specificity will be further discussed in chapter 7.

Sterols

The sterols constitute a third major class of membrane polar lipids in addition to the phospholipids and glycolipids. The most abundant sterol in animal tissues is cholesterol (figure 5.29) which has a polar hydroxyl group at one extreme of a compact hydrophobic molecule (figure 5.30). 7-Dehydrocholesterol (with a second double bond at C-7) is a minor component of some mammalian intracellular membranes.

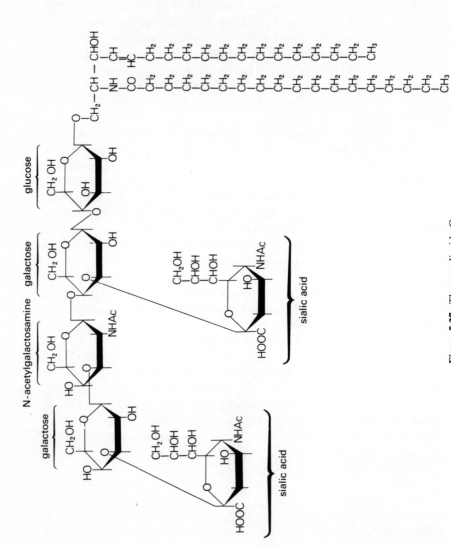

Figure 5.27. The ganglioside G_{D1a}.

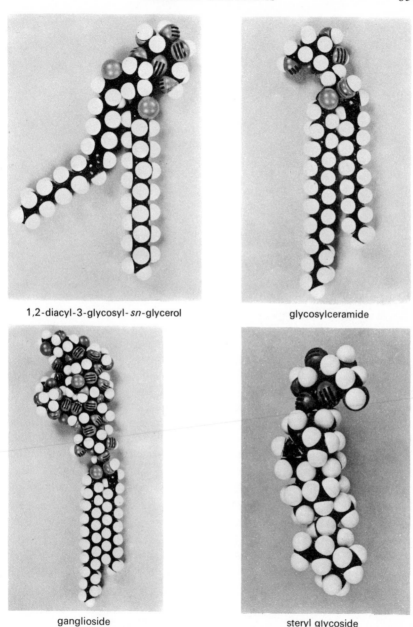

1,2-diacyl-3-glycosyl-*sn*-glycerol

glycosylceramide

ganglioside

steryl glycoside

Figure 5.28. Space-filling models of glycolipids.

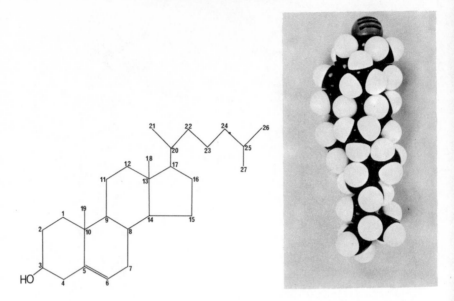

Figure 5.29. Cholesterol.   **Figure 5.30.** Space-filling model of cholesterol.

Cholesterol is known to cause condensation of phospholipid mono-layers (figure 5.31) spread at an air-water interface. It also reduces the permeability of model phospholipid bilayers and 'smooths out' sharp lipid phase transitions, introducing an intermediate fluid state (chapters 6 and 10). In natural membranes it has been shown to act as a stabilizing factor, in that increased osmotic fragility follows cholesterol depletion. All these stabilizing and controlling effects have been attributed to the ability of cholesterol to influence the packing of the hydrocarbon chains of phospholipids, possibly by way of an equimolar cholesterol-phospholipid complex (chapter 6).

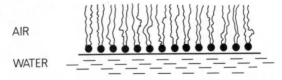

Figure 5.31. Polar lipid monolayer at an air-water interface.

Cholesterol is usually only a minor sterol component of membranes of higher plants in which the major sterols are frequently sitosterol (figure 5.32) and stigmasterol (figure 5.33). These molecules show structural features which are widespread among plant sterols (phytosterols). Phytosterols commonly have an additional side chain at C-24, which may be methylene, methyl, ethylidene or ethyl, and/or a double bond at C-22.

Rather more structural variation is seen among the sterols of eukaryotic micro-organisms, in which a double bond may be present at C-7 in addition to, or instead of, the C-5 double bond in the above structures. Thus ergosterol (figure 5.34) is a major sterol of membranes in such organisms. Ergosterol has been shown to stabilize yeast plasma membranes against osmotic lysis, and similar stabilization can be effected by stigmasterol. Sterols lacking the C-22 double bond are less effective in

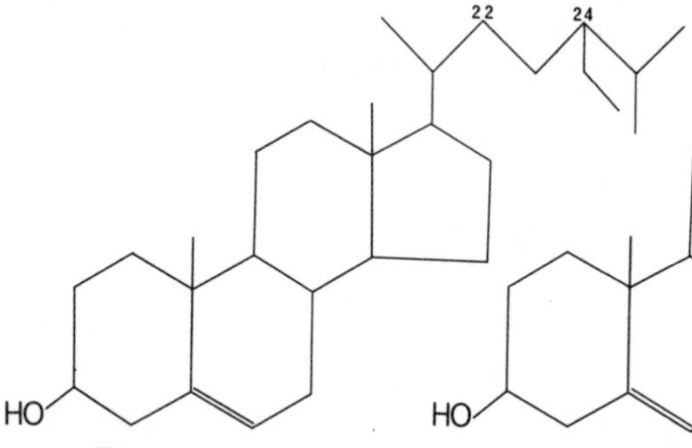

Figure 5.32. Sitosterol.

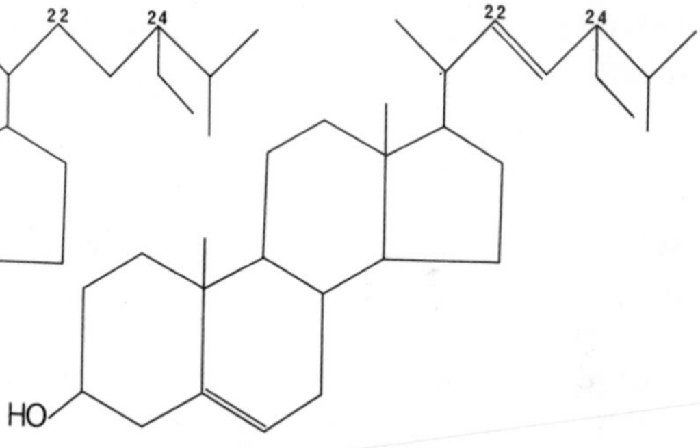

Figure 5.33. Stigmasterol.

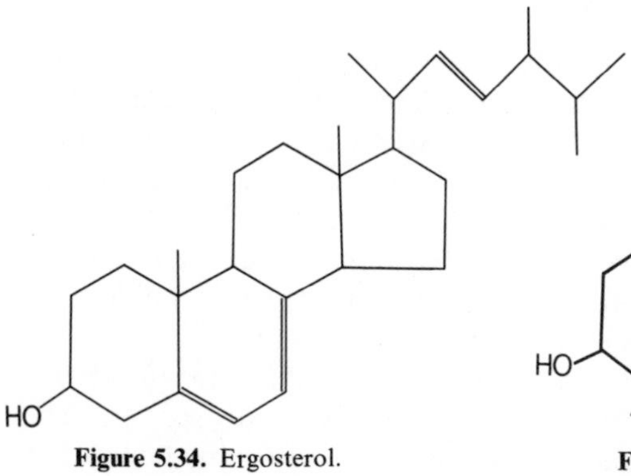

Figure 5.34. Ergosterol.

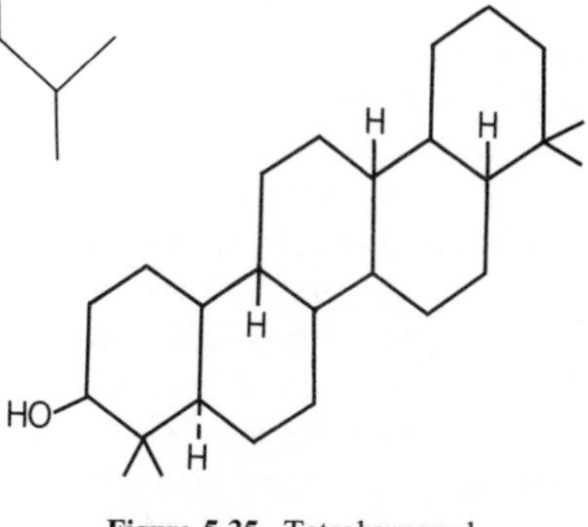

Figure 5.35. Tetrahymanol.

this respect and it may be that the rigidity of an unsaturated C-17 alkyl chain is important in the stabilizing mechanisms.

An interesting sterol-like compound, tetrahymanol (figure 5.35), is the principal neutral lipid of the ciliary membranes of the protozoan *Tetrahymena pyriformis* (page 58) where it presumably performs a stabilizing role similar to that of cholesterol in mammalian membranes.

Other polar lipids in membranes

Membrane polar lipids which do not fall into one of the three major classes of phospholipid, glycolipid and sterols are usually minor components. Mono- and diacylglycerols and free fatty acids, for example, all occur in small amounts in animal membranes. Free fatty acids have been shown to be present in small amounts in brain, where they are associated with the synaptic membranes. The level of free fatty acids varies with the functional state of the membrane, and it has been suggested that they may be involved in changes in membrane permeability.

Non-polar lipids

While our model of a lipid bilayer (figure 5.3) is dependent on amphipathic lipid constituents, biological membranes have been reported to contain minor amounts of non-polar lipids such as triacylglycerol (the lipid of depot fat) and steryl esters of long-chain fatty acids. The location and role of such compounds is not clear, and it is probable that, in many cases, they arise artefactually in the membrane preparations concerned.

Distribution of membrane lipids

Lipid compositions of animal membranes vary both with their tissue source and intracellular location. Apart from plasma membranes, however, which show considerable species variation, corresponding membranes from different vertebrates have remarkably constant lipid composition, although the fatty-acid content of individual lipid components may vary with temperature or nutritional state. Total lipid and phospholipid compositions for subcellular membranes of rat liver are shown in figures 5.36 and 5.37 respectively.

Plasma membranes, as already mentioned, are subject to considerable compositional variation, both with organ and with species. They generally contain most of the glycolipids of the cell and show a high

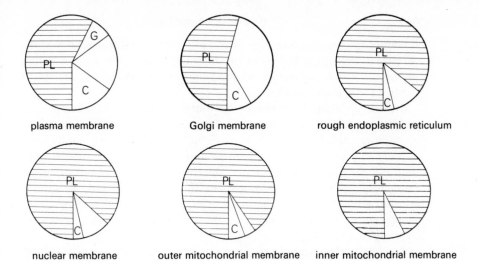

Figure 5.36. Lipid composition by weight of different subcellular membranes of rat liver; PL = phospholipid; C = cholesterol; G = glycolipid; unlabelled = mono-, di- and triacylglycerols, free fatty acids and steryl esters. The high neutral lipid content found for Golgi membrane is probably partly attributable to the presence of very-low-density lipoprotein precursor particles in the Golgi fraction.

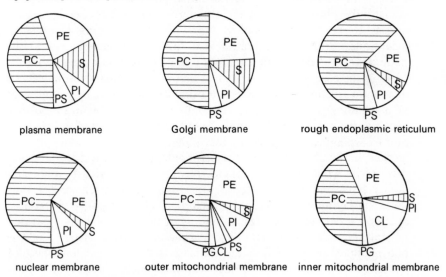

Figure 5.37. Phospholipid composition by weight of different subcellular membranes of rat liver. PC = *sn*-glycero-3-phosphocholines; PE = *sn*-glycero-3-phosphoethanolamines; PI = *sn*-glycero-3-phosphoinositols; PS = *sn*-glycero-3-phosphoserines; CL = cardiolipin [1,3-bis(phosphatidyl)-glycerol]; S = sphingomyelin.

content of sphingomyelin which, with its long-chain saturated fatty acids, will tend to reduce the fluidity of the outer membrane. The relatively high content of cholesterol will at the same time act, as mentioned earlier, to stabilize the cell membrane.

Mitochondrial inner membrane has a high proportion of phospholipid and no (or very little) cholesterol, whereas the content of 1,3-bis(phosphatidyl)-glycerol (cardiolipin) is much higher than in other membranes. In many respects (e.g. cholesterol and sphingomyelin content) the lipid composition of Golgi membrane is intermediate between that of rough endoplasmic reticulum membrane on the one hand and of plasma membrane on the other. This may reflect the proposed role of Golgi membranes as intermediates in the biosynthesis of plasma membranes (chapter 6).

Much less information is available concerning the lipid components of defined plant membrane preparations, apart from those of chloroplasts and mitochondria. Chloroplast membranes differ from other plant and from mammalian membranes in that glycoglycerolipids predominate over phospholipids, and in that the major phospholipid component is phosphatidylglycerol. In so far as other plant membranes have been studied, the phospholipid, if not the sterol (see page 85), compositions appear to be rather similar to those of corresponding mammalian membranes. Fungal membranes also fall into a general eukaryotic pattern which is quite different from that of prokaryotic membranes.

As in the case of plant and fungal tissues, the preparation of defined membrane fractions from bacteria is usually complicated by the presence of a peptidoglycan-containing cell wall. In the case of Gram-positive bacteria, the outer wall is generally assumed to be devoid of lipids which can accordingly be assigned to the underlying cell membrane. This is marked by its content of glycoglycerolipids, particularly the diglycosyl derivatives, while the phospholipid fraction is dominated by phosphatidylglycerol and its derivatives, the O-aminoacyl esters and 1,3-bis (phosphatidyl)-glycerol. The cell envelope of Gram-negative bacteria is, of course, even more complex, consisting of two lipid bilayers separated by the peptidoglycan-containing periplasmic space (page 7). The phospholipids of the bilayers are characterized by the predominance of phosphatidylethanolamine together with its N-methyl and NN-dimethyl derivatives and lesser amounts of phosphatidylglycerol and 1,3-bis (phosphatidyl)-glycerol. Analyses of the separated (see page 56) membranes of E. coli have shown little quantitative difference between the phospholipids of the cytoplasmic and outer membranes. The outer membrane of Gram-negative bacteria also contains a unique lipopoly-saccharide (LPS) typified by that from *Salmonella typhimurium* and shown in figure 5.38. The lipid A portion of LPS replaces part of the phospholipid in the outer half of the lipid bilayer, while the O-specific

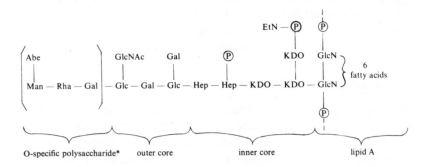

Figure 5.38. A lipopolysaccharide (LPS) from the outer membrane of *Salmonella typhimurium*. Abe = abequose (3-deoxy-D-fucose); Man = D-Mannose; Rha = L-Rhamnose (6-deoxy-L-Mannose); Hep = L-*glycero*-D-*manno*-heptose; KDO = 3-deoxy-D-*manno*-octulosonic acid; EtN = ethanolamine; ℗ = phosphate. LPS normally contains three of the above monomer units linked via pyrophosphate linkages (dotted lines).
  * Three types of surface antigens have been used to classify enteric bacteria: K-antigens (capsular polysaccharides); H-antigens (flagellar proteins) and O-antigens.

polysaccharide chains extend away from the cell and carry serological specificity as well as phage receptor sites in their sugar sequences (cf. chapter 7).

Phosphatidylcholine and phosphatidylinositol are comparatively rare in bacteria. Sterols also are generally absent from prokaryotic micro-organisms apart from mycoplasmas, which, perhaps significantly, do not have cell walls.

Proteins

Membrane functions such as enzymic, transport and receptor activities are believed to be largely mediated by proteins. It might accordingly be expected that the protein content of a particular membrane will reflect the level of activity of that membrane, and this has been found to be generally true. Thus myelin (chapter 3) whose main function is probably that of an insulator contains only 20–30 % protein by weight. Animal cell plasma membranes with approximately 50 % protein, on the other hand, are clearly involved in many enzymic and transport processes, whereas inner mitochondrial membrane with much higher functional activity contains 75 % protein. Such considerations can provide only a rough guide to protein distribution, and often functional requirements necessitate the concentration of protein molecules within specific areas of a membrane (see e.g. page 25) to give local densities that are not necessarily reflected in overall contents of the type quoted above.

Peripheral and integral proteins

Membrane proteins have been classified in two general categories, peripheral and integral, based on their ease of dissociation from the membrane. *Peripheral* proteins (also referred to as *extrinsic*, *external*, or *membrane-associated* proteins) are relatively easily dissociated by mild treatment such as manipulations of ionic strength or pH. These proteins are released in soluble non-aggregated form, free from contaminating lipid, and may be purified by standard techniques of protein fractionation. *Integral* (or *intrinsic*) proteins are in general only isolated by more drastic treatments involving extensive disruption of the membrane by detergents or *chaotropic agents*. Chaotropic agents act by 'destructuring' water, so reducing the unfavourable entropy effects of an apolar residue in structured water. They accordingly tend to disrupt hydrophobic associations (page 65) in membranes. Such agents can be organic molecules (e.g. guanidine hydrochloride, urea) or inorganic ions with large hydration shells (e.g. Li^+, Cl^-, SCN^-). Removal of the solubilizing agent from extracted integral membrane proteins often results in their aggregation and precipitation. Largely because of such difficulties, the separation and characterization of membrane proteins has lagged behind corresponding studies on the lipid components. In many cases, however, procedures have now been developed for the fractionation of integral proteins by standard techniques such as gel electrophoresis, gel filtration and ion-exchange chromatography in the presence of detergents and other disaggregating media. *Affinity chromatography* in similar media has proved to be particularly effective in a number of cases. This technique, summarized in figure 5.39, makes use of the functional activity of the protein and may be used to extract minor components from a protein mixture.

The *proteolipids* are a class of particularly hydrophobic integral membrane proteins operationally defined by Folch-Pi in 1951 as lipoproteins soluble in chloroform-methanol mixtures. Proteolipids have been demonstrated in a wide range of membranes, including those of insect and vertebrate muscle, mitochondria and erythrocytes. The best-characterized of the class are the myelin proteolipids (Folch proteolipids), of which the major protein component, lipophilin, has been purified by

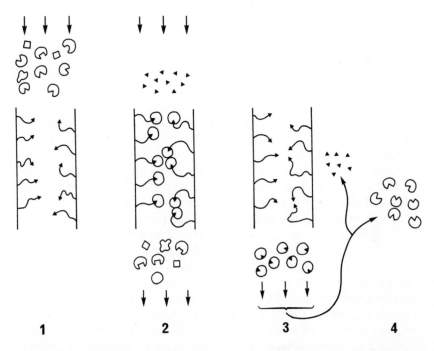

Figure 5.39. Schematic diagram showing the principle of affinity chromatography as applicable to the purification of a membrane protein (e.g. a receptor or enzyme) that specifically binds a ligand (▲). A detergent-soluble mixture of membrane proteins is passed down the column containing the ligand (▲) bound to an insoluble support (1). Proteins that specifically bind the ligand (▲) are retarded on the column, while all other proteins in the mixture are eluted (2). The column is then eluted with excess free ligand (▲) which successfully competes with bound ligand for the receptor protein and so washes the protein off the column (3). The ligand is finally separated (e.g. by dialysis) from purified protein (4).

The same principle can be applied to the purification of glycoproteins by using column-bound lectins (see p. 131) that bind specific carbohydrate components of the required glycoprotein. Elution in this case is with the pure carbohydrate (often a monosaccharide) which displaces pure glycoprotein from the bound lectin.

filtration on hydrophobic gels in chloroform-methanol. Lipophilin contains two molecules of tightly-bound fatty acid per molecule of protein, and is believed to span the lipid bilayer of the myelin membrane.

Integral proteins in general are envisaged as penetrating the hydrophobic interior of the membrane to a greater or lesser extent, whereas peripheral proteins are believed to be associated with the polar head groups on the outer faces of the lipid bilayer. Sections of integral proteins in immediate contact with hydrophobic lipid chains in the bilayer might be expected to be made up largely of non-polar amino acids. A number of proteins, such as cytochrome b_5 and cytochrome b_5 reductase of endoplasmic reticulum and some haemagglutinins of viral membranes, have been shown to be anchored to the lipid bilayer by polypeptide sections composed largely of hydrophobic amino acids. In the case of glycophorin A of human erythrocyte membrane (page 95), the total amino-acid sequence has been determined and a section of 22 non-polar residues is known to cross the membrane. A more complex arrangement occurs in the small-molecular-weight lipoprotein that spans the outer membrane (page 7) of *E. Coli*. Hydrophobic amino acids occur in a repeating sequence so as to give an α-helix having one hydrophobic face. Suitable arrangement of six such helices could provide a superhelix with a hydrophobic outer shell and containing interior hydrophilic channels. Complexes of this type are clearly attractive candidates in a search for transport-mediating proteins (see also page 173).

The presence of particular sequences of hydrophobic amino acids is not generally reflected in the overall amino-acid content of integral membrane proteins. Attempts have been made to demonstrate that integral proteins have a relatively high percentage of 'non-polar' amino acids, but comparisons of this kind are subject to the somewhat arbitrary classification of individual amino acids as 'polar' or 'non-polar'. A more sophisticated approach is to assign a hydrophobicity index to individual amino acids on the basis of their solubility in solvents of different polarity, and so to assess the 'average hydrophobicity' of a given protein. Neither approach has shown consistent differences between integral proteins on the one hand and peripheral or soluble non-membrane proteins on the other. It is likely that hydrophobic surfaces resulting from tertiary and quaternary protein structure would not be detectable by overall amino-acid analysis. Moreover, even continuous intramembrane hydrophobic sections may be relatively short and balanced by externally-disposed hydrophilic sequences, as in the case of glycophorin A (page 95) of erythrocyte membrane.

Glycoproteins

Most cellular membranes contain glycoproteins which have a number of more or less complex oligosaccharide side chains, each attached by a single glycopeptide linkage to the polypeptide backbone. Membrane carbohydrate, both glycoprotein and glycolipid in nature, is asymmetrically distributed, and in general only those proteins that have sections exposed on the extracytoplasmic faces of cellular membranes are glycosylated. This localization of carbohydrate residues is well illustrated in the case of human erythrocyte membrane proteins discussed in the next section. The detailed carbohydrate structures of membrane glycoproteins will be dealt with in chapter 7.

Proteins of the human erythrocyte membrane

Polyacrylamide gel electrophoresis (PAGE) in the presence of the detergent sodium dodecyl sulphate (SDS) is probably the most widely used technique for the analysis of membrane proteins. It is generally assumed that SDS disrupts all non-covalent interactions and binds to the resulting disaggregated polypeptides with a constant weight of SDS per unit length of peptide to give SDS-protein complexes of equivalent shape. In this case electrophoretic mobility should be determined solely by size, and should give a reliable estimate of the molecular weight of each polypeptide component in a mixture. Although the above assumptions have been disputed and in some cases, particularly those of glycoproteins, molecular weights determined by SDS-PAGE are known to be erroneous, the technique has provided a great deal of information concerning the protein composition of biological membranes.

SDS-PAGE analyses of plasma membranes of animal cells commonly show 7–20 proteins which stain with Coomassie blue. A similar range of proteins has been detected in other membranes, such as endoplasmic reticulum and surface membranes of bacterial cells. Mitochondrial and chloroplast membranes are apparently more complex, containing 20–40 species of polypeptide chain, whereas certain specialized membranes may have just one or two protein components. The disc membranes of retinal rods, for example, have been shown to contain only one major protein, rhodopsin, which comprises some 50% by weight of the membrane (chapter 9). The commonly quoted SDS-PAGE Coomassie blue-stained pattern of the human erythrocyte membrane is relatively simple, containing just 7–9 major bands. This is shown in figure 5.40, together with

the simpler pattern obtained when this gel is stained with periodic acid—
Schiff's (PAS) reagent which is fairly specific for carbohydrate. Periodic
acid cleaves the carbon chain between neighbouring hydoxyl groups to
give carbonyl groups, which will react with Schiff's reagent (leuco-
fuchsin) to give a pink or purple Schiff's Base.

The relative simplicity of its protein content is an advantage in the use
of the erythrocyte membrane as a model for studies on membrane
proteins. Other advantages are that the adult mammalian erythrocyte is
an end cell with a single membrane containing a constant population of
protein components, and that it is readily available on a large scale (see
also page 46). There is accordingly considerable information concerning
the proteins of the membrane, and an outline of these data will serve to
give an indication of the types, and particularly the locations, of membrane proteins in general, although it must be borne in mind that the
erythrocyte is not a typical cell.

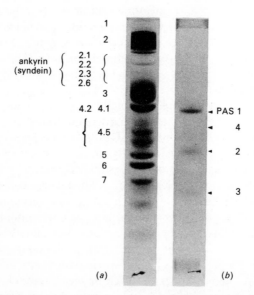

Figure 5.40. SDS-PAGE patterns of human erythrocyte membrane proteins stained with
Coomassie brilliant blue (a) and PAS (b). The gels were run under the conditions described
by G. Fairbanks et al. in *Biochemistry* (1971), **10**, 2606–2617, and the nomenclature is that
of these authors; from H. Furthmayr in *Receptors and Recognition* (ed. P. Cuatrecasas and
M. F. Greaves), Vol. 3, Chapman and Hall, London (1977), by courtesy of Professor H.
Furthmayr, Department of Pathology, University of Yale.

Erythrocytes can be made to swell in hypotonic media when they become 'leaky', losing their haemogloblin and other cytoplasmic constituents (page 46). In this state both internal and external faces of the remaining plasma membrane or *ghost* are accessible to added reagents. Ghosts can be 'resealed' in isotonic sodium chloride solution when the membrane regains the relative impermeability of the original intact erythrocyte. In this state, as in intact erythrocytes, only the external face of the membrane is accessible to subsequently-added reagents. It is accordingly possible to modify selectively proteins exposed at either the outer or the inner surface of the plasma membrane by using radio-chemical labels or enzymic treatment. Subsequent analysis of treated membranes by SDS-PAGE can show which polypeptides have been modified and so indicate the locations of the labelled proteins with respect to the membrane.

As a result of such experimental approaches, a picture of the human erythrocyte membrane has emerged in which the bulk of the major proteins are believed to be peripheral proteins associated exclusively with the cytoplasmic face of the lipid bilayer, whereas two major proteins [band 3 polypeptide and glycophorin A (PAS-2), see figure 5.40] are exposed at both faces of the membrane.

The SDS-PAGE gels shown in figure 5.40 show four major bands when stained with PAS, which is particularly sensitive to the presence of sialic acid. Extraction of the membrane with phenol and the chaotropic agent, lithium diiodosalicylate (LIS), gives a sialoglycoprotein fraction, glycophorin, containing 60% of its dry weight as carbohydrate and including most of the sialic acid of the cell. Glycophorin is a mixture which shows a number of PAS bands on SDS-PAGE. It can be fractionated by gel filtration, lectin-affinity chromatography (figure 5.39), and preparative SDS-PAGE techniques, to give the apparently homogeneous sialoglycopeptides, glycophorins A and B, as well as at least one other glycopeptide. The amino-acid sequence of the major component, glycophorin A, has been determined, and this glycoprotein has in many ways served as a model for membrane glycoprotein structural studies in general (see e.g. pages 151, 212). Although purified glycophorin A appears to be homogeneous (see also page 136), at least in the protein part, it gives two PAS-stained bands (PAS-1 and PAS-2) on SDS-PAGE (figure 5.40). This caused considerable confusion until it was realized that PAS-1 represents a dimeric form of the monomer PAS-2. Glycophorin A is an integral membrane protein with a molecular weight of 31 000. Its single polypeptide chain containing 131 amino acids spans the membrane and can be regarded as being made up of three distinct sections. A hydrophilic N-terminal section extends from the outer surface of the lipid bilayer into the extracellular medium and carries all the carbohydrate of the molecule. This section is rich in hydroxy-amino acids to which 15 small sugar complexes are linked, while a single asparagine residue provides the linkage for a more complex oligosaccharide side chain. The detailed structures and possible functions of the carbohydrate side chains will be discussed in chapter 7. The middle polypeptide section crosses the lipid bilayer, probably in the form of an α-helix and comprises largely hydrophobic amino acids. Finally the C-terminal sequence, rich in proline residues and acidic amino acids, is exposed to the cytoplasm on the inner face of the membrane. A schematic diagram of glycophorin A is shown in figure 5.41. Glycophorin B has been studied less extensively. In this case also, SDS-PAGE patterns have caused confusion in that re-electrophoresis of the apparently pure monomer (PAS-3) under the conditions of figure 5.40 gives more than one PAS-staining band. Some clarification of PAS patterns of erythrocyte membrane polypeptides has been provided by the use of a discontinuous gel system which affords better resolution. SDS-PAGE analysis of normal human erythrocyte membrane (figure 5.42) or of a total LIS-phenol extract on discontinuous gels shows at least seven PAS-staining bands in addition to a faint band corresponding to Coomassie blue band 3 (figure 5.40). Five of these bands [α_2, α/δ, δ_2, α, δ in the nomenclature of Anstee et $al.$ (figure 5.42)] can be assigned to monomers or dimers of glycophorins A (α) and B (δ), while two further bands (β and γ) have been attributed by Furthmayr

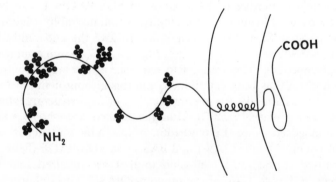

Figure 5.41. Schematic diagram of glycophorin A, the major sialoglycoprotein of the human erythrocyte membrane. ● represents carbohydrate residues.

(figure 5.42) to a third glycophorin, glycophorin C. Designations of the major PAS-staining bands according to different authors are shown in the legend to figure 5.42.

Glycophorin B (band δ or PAS-3) seems to be similar in structure to glycophorin A in that its carbohydrate side chains are all carried on the extracellular N-terminal polypeptide chain and a hydrophobic peptide is immersed in the lipid bilayer. In contrast to glycophorin A, glycophorin B does not seem to be exposed to the cytoplasm on the inner face of the membrane.

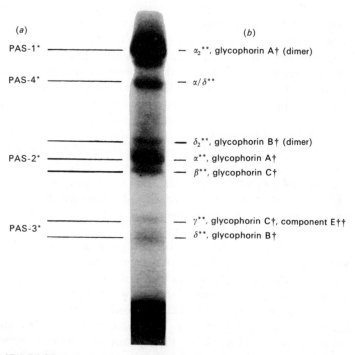

Figure 5.42. SDS-PAGE pattern of human erythrocyte membrane proteins obtained by using the discontinuous electrophoresis system of U.K. Laemmli, *Nature*, (London) (1970), **227**, 680–682 and stained with PAS. Nomenclatures are those applied to (*a*) continuous gels of the type shown in figure 5.39 by *T. L. Steck, *J. Cell. Biol.* (1974), **62**, 1–19, and to (*b*) the discontinuous gel system by **D. J. Anstee, W. J. Mawby and M. J. A. Tanner, *Biochem. J.* (1979), **183**, 193–203, by †H. Furthmayr, *J. Supramol. Struct.* (1978), **9**, 79–95 and by ††W. Dahr, G. Uhlenbruck and H. Knott, *J. Immunogenetics* (1975), **2**, 87–100. The photographic print was kindly supplied by Dr. D. J. Anstee, South West Regional Blood Transfusion Centre, Bristol and Dr. M. J. A. Tanner, Department of Biochemistry, University of Bristol, who also provided a copy of their manuscript prior to its publication.

Band 3 of Coomassie blue-stained SDS-PAGE patterns (figure 5.40) appears as a broad band with a sharp leading edge and a diffuse trailing section encompassing the molecular weight range 90 000–100 000. Present evidence indicates that, while band 3 undoubtedly contains a number of minor components, it primarily comprises a homogeneous polypeptide, *band 3 polypeptide*, which represents about 90 % of the total protein of the band and is present to the extent of some 10^6 copies per cell. Band 3 polypeptide is an integral membrane protein which can be solubilized and purified in the presence of non-ionic detergents (e.g. Triton X-100). It is a glycoprotein containing approximately 8 % by weight of carbohydrate made up largely of galactose, mannose and N-acetylglucosamine, attached via N-glycosidic linkages (page 137) to the polypeptide backbone and exposed to the outside of the membrane. Complete structures of the oligosaccharide side chains have not been reported, but initial results indicate the presence of long, branched chains containing the repeating unit galactose $\beta 1 \rightarrow 4$ N-acetylglucosamine. The chains are apparently heterogeneous even in a relatively pure polypeptide (see page 136), and such heterogeneity could contribute to the broadness of band 3 on SDS-PAGE. The disposition of band 3 polypeptide with respect to the membrane has been the subject of some debate but the present consensus, based on asymmetrically-directed (see page 95) labelling and proteolytic cleavage, is that the polypeptide crosses the lipid bilayer at least once with its glycosylated c-terminal section exposed on the extracellular face of the membrane and its N-terminus within the cytoplasm.

The cytoplasmic portion of band 3 polypeptide appears to bind several peripheral proteins (see below) in a supramolecular complex, the functional significance of which is not understood. Band 3 polypeptide itself can be specifically covalently labelled by inhibitors of anion-transport [e.g. 4,4'-diisothiocyano-2,2'-stilbene disulphonate (DIDS)] which indicates that the polypeptide is associated with this function (chiefly transport of Cl^- and HCO_3^-) in the erythrocyte. Further support for such a role is provided by reconstitution experiments in which incorporation of purified band 3 polypeptide preparations confers anion-transport activity on lipid vesicles (page 169). Two further integral proteins, acetylcholinesterase and the phosphorylated intermediate of Na^+,K^+-activated ATPase (approximately 10^2 copies per cell) have been identified as minor components of the band 3 region of SDS-PAGE.

The identity of polypeptides involved in the transport of glucose across the erythrocyte membrane is not fully clear. Affinity labelling and reconstitution (page 169) experiments have implicated an integral glyco-protein (band 4.5) of apparent molecular weight 55 000 which migrates in the Coomassie blue band 4 region of SDS-PAGE, but there has been a suggestion that band 4.5 could simply represent a proteolytic break-down product of a true band 3 transporter protein.

The later stages of purification of band 4.5 polypeptide from detergent extracts involves separation from band 7 polypeptide, another erythro-cyte membrane protein about which little is known.

Bands 1, 2, 2.1, etc. (ankyrin), 4.1, 4.2, 5 and 6 account for approxi-mately 40% of the total membrane protein, and available evidence indicates that they represent peripheral proteins located only on the inner face of the membrane. *Spectrin* is the name given to two high-molecular-weight (200 000–250 000) polypeptides (bands 1 and 2) which are readily extracted from the membranes by low ionic strength buffers containing chelating agents (e.g. EDTA). The two polypeptides are not easily separated from each other and have been shown to associate side-by-side into long (100 nm) flexible heterodimers (band 1 plus band 2) which can undergo further end-to-end dimerization into tetramers. Spectrin is extracted from the membrane together with band 5 polypeptide, *actin* (45 000) to which it may be linked by polypeptide 4.1, and it may be that a spectrin-actin network on the cytoplasmic face of the membrane is responsible for maintaining the characteristic biconcave shape of the erythrocyte. It appears that spectrin is also linked to integral membrane proteins such as glycophorin (chapter 6) and it has been suggested that spectrin and actin combine in an actomyosin-like contractile apparatus directing the movement of integral proteins within the plane of the membrane. Support for this suggestion has been provided by the demonstration that antibodies directed against smooth muscle myosin display a weak cross-reactivity with human spectrin, but there are many differences between spectrin and myosin and a contractile role for spectrin remains to be proven.

Band 6 protein (figure 5.40) is glyceraldehyde 3-phosphate dehydro-genase which, together with another glycolytic enzyme, aldolase, and band 4.2 is associated with the cytoplasmic tail of band 3 polypeptide. The latter is thought to be linked to spectrin by the ankyrin polypeptides (2.1, etc.). These associations and also those involving the spectrin-actin system and glycophorin (chapter 6) are shown schematically in figure 5.43.

Our picture of the proteins of the human erythrocyte membrane is far from complete. Apart from many uncertainties concerning the major proteins discussed above, particularly with regard to their functions,

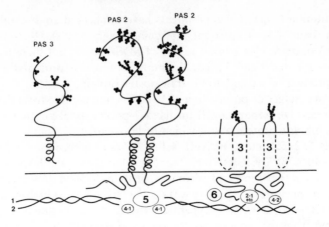

Figure 5.43. Schematic representation of the major polypeptides of the human erythrocyte membrane. Numbering of the polypeptides corresponds to their positions on the SDS-PAGE system of figure 5.40. The dashed sections of band 3 polypeptide are intended to indicate that the chain could cross the lipid bilayer several times. This polypeptide is shown as a dimer in view of its ready cross-linking *in situ* by chemical reagents. Evidence for a similar association of PAS 2 in the membrane is less clear.

many minor proteins undoubtedly exist which are either unlabelled in figure 5.40 or are present in quantities too low to be detected by the resolution and staining of that electrophoretic system. Recent applications of higher-resolution methods to the human erythrocyte membrane dramatically emphasize this point. Isoelectric focusing of an SDS-solubilized membrane extract, followed by PAGE in a second dimension at right angles to the first, separated over two hundred polypeptide components. It is likely that the application of similar techniques to membranes generally will demonstrate just how little we currently appreciate their true complexity.

SUMMARY

1. Approximately 40% of the dry weight of membranes is composed of *lipids*, most of which are *amphipathic* lipids containing hydrophilic and hydrophobic sections linked through a bridging moiety. The bridge may consist of a glycerol molecule, a sphingoid, or the body of a sterol molecule. The amphipathic lipids self-assemble to form a *lipid bilayer* which is the fundamental structure of biological membranes.

2. *Phospholipids* are polar lipids in which the hydrophilic head group consists of one of a range of complex alcohols joined by a phosphodiester linkage to the terminal hydroxy group of a substituted glycerol or sphingoid molecule.

3. The hydrophilic head groups of *glycolipids* consist of mono- or oligo-saccharide units glycosidically-linked directly to the terminal hydroxy-group of one of the three types of bridging moiety in (1).

4. There is more information concerning the lipid composition of animal than of plant or microbial membranes. Within the same animal, corresponding membranes from different organs have different lipid compositions, as do different subcellular membranes from the same cell. Corresponding membranes from different vertebrates generally have similar lipid compositions.

5. Some 60% of the dry weight of membranes is made up of *protein*, which can be classified as *integral* or *peripheral* depending on its ease of dissociation from the membrane. Integral proteins are believed to be intercalated in the lipid bilayer of the membrane.

6. Human *erythrocyte membranes* contain 7–9 major protein components, most of which are peripheral proteins associated with the cytoplasmic face of the membrane. Two integral glycoproteins, *glycophorin* (of which there are at least three types) and *band 3 polypeptide* cross the lipid bilayer and have been studied in some detail.

CHAPTER SIX

STRUCTURAL ORGANIZATION

IN OUR CONSIDERATIONS OF THE CHEMICAL COMPONENTS OF MEMBRANES we have partially anticipated discussion of their organization in the membrane. Thus, the amphipathic nature of membrane lipids was stressed as being compatible with their arrangement as a bilayer (figure 5.3) and in the present chapter we shall be largely concerned with models of membrane structure which incorporate a lipid bilayer as their primary feature.

The Davson-Danielli-Robertson model

Gorter and Grendel in 1925 reported that lipids extracted from erythrocyte membranes spread as a monolayer at an air-water interface (figure 5.31) so as to occupy an area which was approximately twice the total surface area of the intact erythrocytes. They pointed out that this is consistent with the presence in the erythrocyte membrane of a lipid bilayer, in which the hydrocarbon chains all occupy the centre of the bilayer and the polar head groups face outward (cf. figure 5.3). In fact the figures of Gorter and Grendel arose as a result of two errors. Not only were the erythrocyte lipids incompletely extracted, but the erythrocyte surface area was underestimated. These errors partially cancelled each other, and subsequent work has shown that sufficient lipid is indeed available to cover the erythrocyte surface completely as a bilayer, but only if that bilayer is highly expanded.

Apparently independently, Danielli and Davson in 1935 proposed the presence of a lipid bilayer in membranes and, in order to explain what seemed to be anomalously low surface tensions of biological membranes compared with model lipid systems, postulated that the lipid core of natural membranes was sandwiched between two layers of protein. Although this conclusion also was later found to have been based on a misconception, in that phospholipids alone can produce the low surface

102

tensions shown by natural membranes, the Davson-Danielli model dominated discussion of membrane structure for over thirty years. Following the development of electron microscopy, it was found that biological membranes from many sources show a typical three-layered structure (figure 3.1) which can be interpreted in terms of the Davson-Danielli protein-lipid-protein sandwich model. The widespread occur-

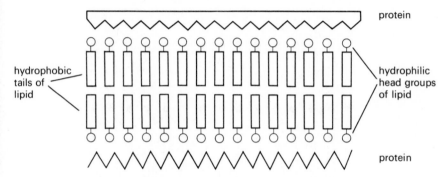

Figure 6.1. Schematic diagram of the Davson-Danielli-Robertson model: from J. D. Robertson, *Ann. N.Y. Acad. Sci.* (1966), **137**, 421–440.

rence of such apparently similar membranes led Robertson in the late 1950s and early 1960s to promote the concept of a universal *unit membrane* based on the Davson-Danielli model. The Davson-Danielli-Robertson model was supported largely by electron microscopic and X-ray diffraction evidence and, as can be seen in figure 6.1, suggested an asymmetric distribution of protein about the lipid core.

Evidence for a lipid bilayer in membranes

The fact that sufficient lipid is available in membranes to form a continuous, if expanded, lipid bilayer has already been mentioned. It has been known for many years that small molecules with a high oil-water partition coefficient penetrate cell membranes most readily, and this is clearly consistent with the idea that the permeability barrier of the cell is a hydrophobic barrier formed by lipid (chapter 1).

The lipid bilayer is thermodynamically stable and is the major structural unit of phospholipid-water systems. Above the transition temperature (see page 115) phosphatidylcholine with low water content exists in a lamellar phase (figure 6.2) consisting of stacked bilayers containing water between the lipid bilayers. As the water content is increased from 0 to 40% (by weight), the width of the aqueous layers increases, and the system remains homogeneous.

Above 40% water a two-phase system results in which maximally-hydrated multilamellar vesicles (*liposomes*) are dispersed in water (figure

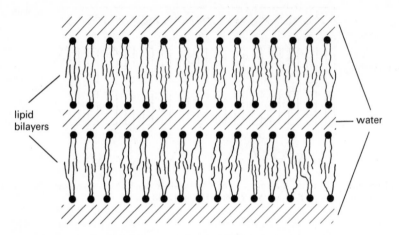

Figure 6.2. Bulk lamellar phase of phosphatidylcholine-water system containing less than 40% water above its transition temperature.

6.3). Such vesicles can be produced simply by shaking phosphatidyl-choline in water, and tend to precipitate on standing. More stable homogeneous liposome preparations can be obtained by ultrasonication, which promotes formation of single-bilayer-bounded vesicles (figure 6.4). These have been shown by many methods, including nuclear magnetic resonance (n.m.r.) and electron spin resonance (e.s.r.) (chapter 10) to be completely sealed. They accordingly constitute a simple model system for permeability studies of lipid bilayers, particularly as swelling of the vesicles can be followed by measurements of light scattering. The water and small-molecule permeabilities of a number of phospholipid and phospholipid-cholesterol systems have been examined in this way and found to approximate to those of natural membranes. Cholesterol was found to decrease liposome permeability in a manner which parallels its ability to condense phospholipid monolayers (chapter 5).

Figure 6.3. Multilamellar vesicles of phosphatidylcholine dispersed in excess water. (a) Diagrammatic representation showing only three concentric bilayers. (b) Freeze-fracture electron micrograph (see p. 110) showing a number of successive bilayers A: from H. Hauser *et al.*, *J. Mol. Biol.* (1970) **53**, 419–433 by courtesy of Dr. E. G. Finer, Unilever Research, Welwyn.

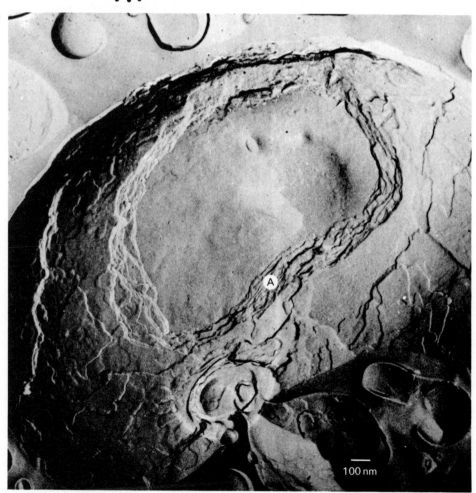

Figure 6.4. Diagrammatic representation of single-bilayer-bounded vesicles obtained by ultrasonication of phosphatidylcholine in water.

A convenient model system for studying the electrical properties of lipid bilayers is the *black lipid film* (figure 6.5). A dilute solution of lipid in a hydrocarbon solvent is applied across a hole (about 2 mm diameter) separating two electrically-insulated compartments filled with electrolyte. Under reflected light and at low magnification, the lipid layer can be observed to thin over a period of about 15 minutes, until it is reduced to bilayer thickness over most of the aperture. At this stage light reflected from the front and back surfaces of the film is very nearly in counter phase, and the film appears to be black. The capacitances and permeabilities of black films of phospholipid and phospholipid-cholesterol mixtures have been measured and found to agree well with values for some biological membranes. Although the electrical resistance of black

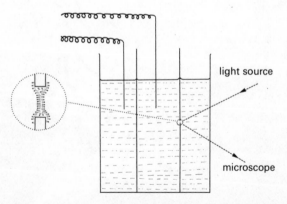

Figure 6.5. Simple apparatus for the study of black lipid films.

lipid films is generally very much higher than that of natural membranes, the resistance may be greatly reduced following incorporation of certain proteins into the lipid film. A particularly interesting observation is that cation and anion selectivities can be induced in both black films and natural membranes by addition of certain cyclic polypeptide antibiotics (chapter 8).

Incorporation of cholesterol into black lipid films stabilizes the film and reduces water permeability, consistent with its ability to condense phospholipid monolayers (chapter 5) and to reduce the fluidity of hydrocarbon chains (see page 116). Cholesterol also reduces the effect of the ion-carrying antibiotics, probably by inhibiting their diffusion in the bilayer.

Distribution of protein in membranes

There is clearly considerable support for the presence of a lipid bilayer in membranes. Evidence for the distribution of protein as depicted in the Davson-Danielli-Robertson model is less convincing. Electron micrographs of a wide variety of membranes show a typical 'tramline' appearance (figure 3.1) which directly suggests the trilaminar structure of the Davson-Danielli-Robertson model. The true significance of the staining patterns observed after $KMnO_4$ or OsO_4 fixation (commonly used in electron microscopy) is not, however, clear. Whereas the three layers seen in electron micrographs could be consistent with the protein-lipid-protein structure, they might well simply reflect the hydrophilic-hydrophobic-hydrophilic sequence across a lipid bilayer. Indeed electron micrographs have been obtained from artificial bilayers which are remarkably similar to those of natural membranes (figure 6.6).

Much of the experimental support for the Davson-Danielli-Robertson model was derived from X-ray diffraction data of myelin. The myelin sheath forms an insulating coat around nerve axons and is produced by a spiral wrapping of many layers of plasma membrane from the Schwann cell (chapter 3). The trilaminar appearance of the Schwann cell membrane and its continuity with the myelin spiral can be clearly seen in electron micrographs of a nerve fibre in the process of myelination (figure 3.13). The electron density distribution across the width of one plasma membrane in the myelin sheath can be obtained from X-ray diffraction patterns, and is made up of two fairly broad maxima separated by a trough (figure 10.21). Such patterns can be explained on the basis of the Davson-Danielli-Robertson model, when the electron-dense outer humps would arise largely from the protein layers which are relatively rich in the heavier O and N atoms. The central trough would represent the light C and H atoms of the hydrocarbon chains. Much the same distribution is also given by a simple bilayer of phospholipid (figure 10.20) and the X-ray diffraction patterns of myelin do not necessarily tell us much about the protein of the membrane.

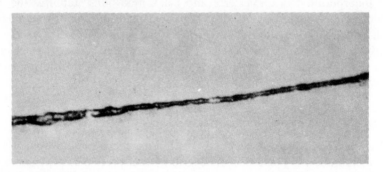

Figure 6.6. Electron micrograph of an artificial bilayer formed from phosphatidylethanolamine in n-decane. The bilayer was fixed with $La(NO_3)_3$ and $KMnO_4$ and dehydrated by air-drying at room temperature: from J. M. Palmier and D. O. Hall, *Progr. Biophys. Mol. Biol.* (1972), **24**, 127–176, by courtesy of Professor J. W. Greenawalt, School of Medicine, The Johns Hopkins University, Baltimore.

The Davson-Danielli-Robertson model envisaged a central lipid bilayer of some 4·0 nm thickness sandwiched between two protein layers, each approximately 2·0 nm thick, so making a total membrane width of about 7·5 nm as determined from electron micrographs of $KMnO_4$-stained membranes. Membrane dimensions vary both with their source and the means of measurement, but the space generally available for protein on the basis of the Davson-Danielli-Robertson model is such that a significant proportion of the membrane protein could be expected to be in the extended β-pleated sheet structure. This structure has ionic side-chains in contact with both the polar lipid head groups of the bilayer and with the aqueous environment outside the membrane. In such a system the ionic groups at the lipid-protein interface are effectively buried in a non-polar environment, without access to the aqueous medium. Moreover a significant proportion of the non-polar residues in the protein must be exposed to the aqueous surroundings. A system of much lower free energy would result if the hydrophobic groupings could all be gathered together in a non-polar environment, leaving as many of the polar residues as possible in direct contact with water.

The mosaic model

Such a situation obtains in the *mosaic model*, forms of which were proposed in the mid 1960s by Singer and by Wallach. This model is illustrated in figure 6.7 where it can be seen to involve globular protein embedded in, and occasionally crossing, a lipid bilayer core. In such a structure the polar groups of lipid and protein are in direct contact with

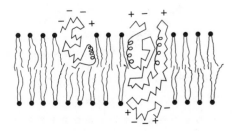

Figure 6.7. The mosaic model of membrane structure. The integral proteins are shown as globular molecules partially embedded in the lipid bilayer. The protruding parts carry the ionic residues of the protein, while the apolar residues are largely in the embedded parts: from S. J. Singer and G. L. Nicolson, *Science* (1972), **175**, 720–731.

the aqueous surroundings, while the non-polar residues of both molecular species are sequestered in the heart of the structure, away from water.

Apart from thermodynamic considerations, the mosaic model is attractive in that it offers a potential route through the lipid bilayer barrier by way of the transmembrane protein. No such provision had been made in the original Davson-Danielli concept, although Danielli did modify his model in 1958 to include protein-lined pores which spanned the membrane.

Evidence for the mosaic model

If the mosaic model is conceptually more attractive than that of Davson, Danielli and Robertson, how does it fit experimental facts? Both models are based on the lipid bilayer for which there is ample evidence. It is in the disposition of membrane protein that the mosaic model differs from its predecessor, and the weakness of some of the experimental support for this aspect of the Davson, Danielli, Robertson model has already been discussed. The mosaic model requires that some protein molecules be embedded in the lipid bilayer. Now the existence of 'integral' membrane proteins was discussed in chapter 5, in which they were operationally defined as a class of proteins that could only be extracted by agents causing extensive disruption of the membrane. This property of integral proteins is consistent with their immersion to a greater or lesser extent in

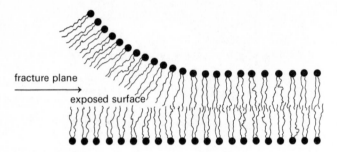

fracture plane

exposed surface

Figure 6.8. Freeze-fracture of a lipid bilayer showing the plane of fracture and the exposed surface.

the lipid bilayer, an arrangement that is further supported by discoveries of localized hydrophobic sections in the proteins (pages 92, 96). Evidence that some proteins actually cross the lipid bilayer is provided by SDS electrophoretic patterns, in which individual polypeptides (e.g. glycophorin A and band 3 polypeptide of the erythrocyte) can be seen to be accessible to reagents at both faces of the membrane. Perhaps the most convincing demonstration of the existence of proteins embedded in the membrane comes from *freeze-fracture electron microscopy*. In this procedure, a fresh membrane specimen is rapidly frozen under vacuum and fractured with a microtome knife; this cleaves the membrane along its interior hydrophobic face (figure 6.8). Frozen water is then sublimed (*freeze-etched*) from the exposed surface, which is metal shadowed and replicated. Electron microscopy of the surface replica reveals the topography of the interior hydrophobic exposed face. A characteristic feature of the cleaved surface of most membranes studied in this way is a smooth matrix interrupted by a large number of fairly uniform particles (figures 6.9 and 6.10). These particles are attributed to integral proteins which, if they cross the mid-plane of the membrane, will remain as part

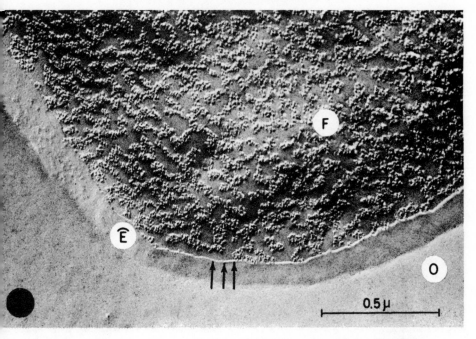

Figure 6.9. Freeze-fracture electron micrograph of an erythrocyte membrane. The fracture plane F is bounded by the arrows and can be seen to be covered with particles of approximate diameter 8·5 nm: from P. Pinto da Silva and D. Branton. *J. Cell. Biol.* (1970), **45**, 598–605, by courtesy of Professor D. Branton. The Biological Laboratories, Harvard University.

nature of the particles in the erythrocyte membrane has been provided by lectin (chapter 7) labelling studies in which the distribution of band 3 glycoprotein (figure 6.12) and also glycophorin on the outer surface of the membrane was found to correspond to the distribution pattern of the intramembrane particles. More convincingly, freeze-fracture of a reconstituted complex formed by introduction of purified band 3 into an artificial lipid bilayer gave a pattern of intramembrane particles indistinguishable from that on the native membrane. Freeze-fracture of protein-free lipid bilayers generally gives particle-free fracture faces. The presence of glycophorin as well as band 3 polypeptide in the erythrocyte intramembrane particles is less certain. In fact, human erythrocytes of blood group En(a-), a rare homozygous condition in which glycophorin A is completely absent, show freeze-fracture particles identical in number and morphology with those of normal cells.

Figure 6.10. Freeze-fracture electron micrograph of chloroplast lamellae. The exposed surface of the fracture face A is covered with particles 16–20 nm in diameter: from D. Branton and R. B. Park, *J. Ultrastruct. Res.* (1967), **19,** 283–303, by courtesy of Professor D. Branton, The Biological Laboratories, Harvard University.

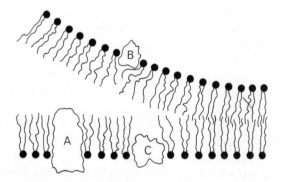

Figure 6.11. Illustration of the exposure of globular intrinsic protein by freeze-fracture of a membrane lipid bilayer. Only protein A, which crosses the mid-plane of the membrane, will be associated with the exposed surface.

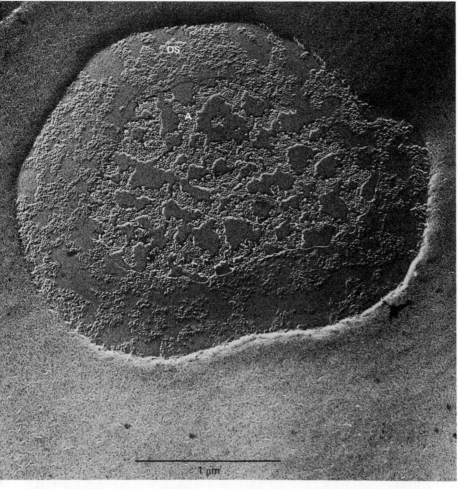

Figure 6.12. Freeze-fracture electron micrograph of human erythrocyte membrane labelled on its outer surface with ferritin-concanavalin A. The ghosts were first incubated in the presence of Ca^{++} to aggregate the membrane intercalated particles. Incubation with ferritin-concanavalin A (chapter 7) then specifically labelled the Band 3 glycoprotein on the outer membrane surface (glycophorin interaction with concanavalin A is known to be negligible). Freeze-fracture, etching and metal-shadowing gave rise to the above electron micrograph in which it can be seen that the pattern of aggregation of the label on the outer surface (OS) closely resembles the random network of aggregated membrane intercalated particles on the freeze-fracture face (A). The pattern of distribution of the ferritin-concanavalin A is continuous and contiguous with that of the membrane intercalated particles: from P. Pinto da Silva and G. Nicolson, *Biochim. Biophys. Acta* (1974), **363**, 311–319, by courtesy of Dr. P. Pinta da Silva, The Salk Institute for Biological Sciences, San Diego.

As discussed in chapter 5, not all membrane proteins are integral proteins, and the mosaic model shown in figure 6.7 should be extended to account for discrete peripheral proteins associated via electrostatic interactions with the hydrophilic head groups of the lipid bilayer.

The fluid mosaic (Singer-Nicolson) model

The mosaic model was amplified by Singer and Nicolson in 1972 when they stressed the dynamic aspects of membrane structure in describing the *fluid mosaic model*. The refined model was most aptly illustrated in their now classic diagram (figure 6.13) in which globular proteins are shown to resemble 'icebergs' floating in a 'sea' of lipid bilayer. This early picture stressed the potential mobility of membrane integral proteins in the plane of the bilayer, and during the mid-1970s the concept stimulated many investigations into the lateral diffusion properties of membrane

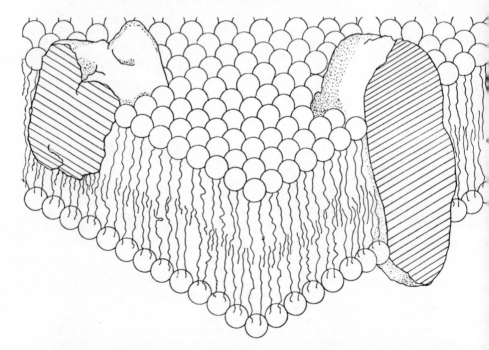

Figure 6.13. The fluid mosaic model. The globular proteins are shown as immersed in and occasionally crossing a lipid bilayer matrix (note that sterols are not distinguished from phospholipids): from S. J. Singer and G. L. Nicolson, *Science* (1972), **175**, 720–731.

components. More recently it has become increasingly apparent that lateral motions of integral proteins particularly are subject to control mechanisms in which peripheral membrane proteins (chapter 5) and cytoplasmic elements may play a major role. Accordingly modern views of, at least, the plasma membrane are better reflected in more complex illustrations of the type shown in figure 6.16, some aspects of which are discussed in the present section.

The mobility of integral proteins in the membrane arises from the fluidity of the lipid matrix, the hydrocarbon chains of which are largely highly mobile at physiological temperatures. At lower temperatures, hydrated phospholipids exist as *gels* containing crystalline hydrocarbon chain regions (figure 6.14) in which parallel fully-extended hydrocarbon chains tilt away from the perpendicular to the plane of the membrane at an angle which depends upon the degree of hydration (chapter 10). Such chains are largely in the transplanar conformation (figure 5.4a) and undergo only slight torsional oscillations.

As the temperature of a pure phospholipid is raised, molecular motion of the hydrocarbon chains gradually increases until, at a characteristic *transition temperature*, a sharp rise in heat absorption occurs and the mobility of the hydrocarbon chains abruptly increases, giving rise to the *liquid crystalline* state. This gel-to-liquid crystalline phase transition can be followed by a number of physical techniques and the molecular motions of the hydrocarbon chains have been examined in some detail

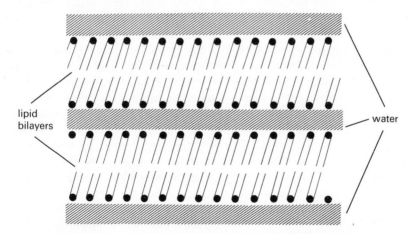

Figure 6.14. Gel phase of hydrated phospholipid below its transition temperature.

by n.m.r. and e.s.r. (chapter 10). As the temperature of the system is raised through that of the phase transition, the hydrocarbon chains rapidly change from the trans-planar conformation of the gel-phase to a mobile state involving considerable flexing and twisting, resulting from carbon-carbon bond rotation. Although interpretations of the liquid crystalline state based on n.m.r. and e.s.r. differ in detail, data from both techniques agree in describing relatively greater mobility for those sections of the hydrocarbon chains furthest removed from the polar head group, i.e. in the hydrophobic centre of the bilayer. The phase transition is also reflected in the mobility of the polar head groups of phospholipid bilayers, although in this case the change in mobility is rather more gradual.

The transition temperature of a given lipid varies with length and degree of unsaturation of the hydrocarbon chains (page 268), while mixtures of lipids show more or less complex phase behaviour which can involve the co-existence of gel and liquid crystalline phases over broad temperature ranges. The effect of cholesterol and other sterols on phospholipid phase transitions is of particular relevance to biological membranes, which commonly contain sterol:phospholipid molar ratios of $(0.5-1):1$. 50% cholesterol (on a molar basis) has the effect of smoothing out or even abolishing the abrupt changes in heat flow (page 271) which characterize the phase transitions of pure phospholipid-water systems. Such behaviour is consistent with the effect of cholesterol of increasing the disorder of phospholipid hydrocarbon chains below the phase transition temperature and of increasing their order above this temperature. The resulting intermediate fluid condition has been attributed to the formation of a 1:1 molecular complex between cholesterol and phospholipid, in which the first ten carbons of the phospholipid tails are particularly restricted by association with the sterol molecule in the bilayer. The exact nature of this association remains a subject for debate.

Natural membranes contain a complex mixture of lipids and many have a high content of sterols, in view of which it is not surprising that their phase transitions are not as sharp as those of synthetic phospholipids, and in many cases are not detectable at all. The hydrocarbon chains in natural membranes are, however, believed to be generally in a fluid state at physiological temperatures, although the presence of sterols may lead to local variations of mobility within the membrane. There is evidence that protein also can influence the mobility of neighbouring lipid molecules. A number of membrane-bound enzymes [e.g. Ca^{++}-dependent ATPase of sarcoplasmic reticulum] have been studied by e.s.r. and other techniques and have been shown to be surrounded by an annulus of immobilized *boundary lipid*, the local phase transitions of which are reflected in the activity of the enzyme (see chapter 8). It should be emphasized that 'immobilized' used in this context means that boundary lipid has relatively little rotational freedom and does not necessarily preclude rapid exchange with surrounding lipid molecules.

As much of the membrane is in a fluid state (the viscosity of biological membranes has been estimated to be that of light machine oil) integral proteins and individual lipids may be expected to show some freedom of lateral motion. E.s.r., n.m.r. and, more recently, fluorescence techniques (chapter 10) have all been applied to measure the lateral diffusion coefficients (D) of lipid probes, and similar values of D (of the order of $10^{-8} cm^2/s$) have been found both in a range of cell membranes and, above the phase transition, in model phospholipid bilayer systems. The lateral motion of integral proteins was first demonstrated by Frye and Edidin in 1970. These workers labelled antigenic integral proteins of human and mouse cells in culture with different fluorescent antibodies. Fusion of the two cell types was then induced by Sendai virus to give compound 'super-cells'. Shortly after fusion, the human and mouse antigenic components were largely segregated in different halves of the fused cell membrane. After incubation at 37 °C for 40 minutes the components were essentially completely intermixed. The mixing process was shown to be independent of chemical energy in the form of ATP, and was attributed to lateral diffusion of the protein molecules in the lipid bilayer (estimated $D = 2 \times 10^{-10} cm^2/s$). The lateral diffusion of protein molecules in a number of cell membranes has since been examined by a variety of methods (see e.g. page 260) and has been found to be generally slower ($D = 10^{-11} \rightarrow 10^{-9} cm^2/s$) than that of lipids, as might be expected from considerations of size. There is considerable evidence that certain membrane proteins are, at times, still less mobile than these figures suggest, and are subject to restraints which limit and control their movement within the plane of the membrane. The distribution of cell surface receptors, for instance, is often non-random, and local associations of membrane proteins (e.g. in cell junctions and mitochondrial multienzyme complexes) frequently occur. Mobility restraints of these kinds are clearly important and need to be considered in any overall picture of membrane dynamics. Some factors which might limit protein mobility have already been mentioned. The existence of sterol-lipid or protein-lipid complexes in the lipid bilayer of the membrane may restrict the area accessible to wandering protein molecules. At temperatures within the broad phase transitions of biological membranes, phase separation of 'liquid' and 'fluid' domains can be demonstrated, and under these conditions freeze-fracture particles are generally concentrated in the 'fluid' domains. Although phase transitions are normally complete well below physiological temperatures, it is conceivable that small crystalline regions might persist at growth temperatures and could exclude (or even sequester) protein membrane components.

It now appears likely that the free lateral diffusion of integral membrane proteins can be limited not only by general factors of the type discussed above but also by specific interaction with other membrane proteins. This aspect of mobility control is well illustrated by reference to the erythrocyte membrane.

The lateral mobility of erythrocyte membrane integral proteins is relatively restricted (D = approximately $4 \times 10^{-11} \mathrm{cm}^2/\mathrm{s}$ at $37\,^{\circ}\mathrm{C}$) compared to similar proteins on other cells, and this has been attributed to their association with peripheral proteins on the cytoplasmic face of the membrane. Aggregation by specific antibody of the peripheral protein spectrin (page 99) in erythrocyte ghosts leads to aggregation of glyco-phorin-bound sialic acid on the outer surface of the membrane. Conversely, when the integral glycoproteins, glycophorin and band 3 polypeptides are made to aggregate (see figure 6.12) by multivalent lectins (chapter 7) specific for carbohydrate residues on their extracellular parts, dimethylmalonimidate-induced chemical cross-linking of spectrin is enhanced, suggesting a concomitant rearrangement of this cytoplasmic peripheral protein. It appears, therefore, that lateral movement of the transmembrane proteins is linked to spectrin on the inner face of the membrane. This idea is further supported by recent reports of the purification from spectrin-depleted erythrocyte ghosts of a water-soluble proteolysis fragment (molecular weight 72 000) that could represent the membrane attachment site for spectrin. The 72 000 fragment binds spectrin, competitively inhibits binding of spectrin to spectrin-depleted vesicles, and can be used to cause rapid and complete dissociation of spectrin from non-depleted vesicles. When spectrin is selectively removed from erythrocyte membranes in this way, the lateral mobility of integral proteins is increased. Although the origin of the 72 000 fragment is not certain, similarities have been found between it and components of the ankyrin complex (figure 5.40) and it is possible that some or all of these components link band 3 polypeptide to spectrin. It has been suggested (see also page 99) that spectrin is not just linked to integral membrane proteins but combines with actin in a contractile apparatus to control their lateral movement in the plane of the membrane. Whereas evidence for such a contractile system in the erythrocyte is currently slight, the idea is attractive in providing a parallel with the postulated role of cytoskeletal elements in other eukaryotic cells.

It has already been stressed that the erythrocyte is not a typical mammalian cell; in fact, spectrin has not been identified elsewhere. Nonetheless most eukaryotic cells contain myosin-like molecules which, together with actin, are involved with a membrane-associated *cytoskeletal* system. Cytoskeletal components include *microfilaments*, which probably contain actin, *thick filaments* similar to polymerized myosin, and *microtubules* which are polymeric complexes of the protein tubulin. Microfilaments and microtubules are disrupted by cytochalasin B (a fungal metabolite) and colchicine (an alkaloid drug) respectively and, largely on the basis of the effects of these agents, microfilaments and microtubules have been implicated in the control of surface receptors on lymphocytes, and other mammalian cells. Lymphocyte surfaces carry interconnecting networks of immunoglobulin (Ig) molecules which upon binding of anti-Ig antibodies, aggregate into 'clusters' and 'patches' and subsequently form 'caps' at one end of the cell (figure 6.15). 'Capping' can be inhibited by cytochalasin B, whereas colchicine has little effect, or even enhances capping. The combined effect of cytochalasin B and colchicine, on the other hand, is to hinder capping in a striking manner. Patching and capping of Ig by anti-Ig has been reported on both T (thymus-derived) and B (bone-marrow derived) lymphocytes from a range of species, and similar sequences of events have been induced by multivalent ligands specific for a variety of surface receptors on both lymphocytes and fibroblasts. As a result of the effects of cytochalasin, colchicine and

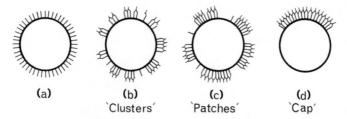

(a)	(b)	(c)	(d)
	'Clusters'	'Patches'	'Cap'

Figure 6.15. Stages resulting from the interaction of a multivalent ligand with surface receptors on a lymphoid cell. The receptors are initially dispersed in an interconnecting network over the cell surface (*a*). On addition of the multivalent ligand (λ) at physiological temperatures the receptors aggregate into 'clusters' (*b*) and 'patches' (*c*) and subsequently form 'caps' (*d*) at one end of the cell.

120   STRUCTURAL ORGANIZATION

related drugs on receptor redistribution, dual and opposing roles for microfilaments and microtubules have been proposed in the control of transmembrane receptors. These proposals are summarized in figure 6.16 in which microfilaments (shown as actomyosin complexes) and microtubules are seen to be linked to each other and to plasma membrane components at specific 'nucleation' points. The function of microfilaments is envisaged as being mainly contractile, serving to direct the movement of receptors within the plane of the membrane, whereas that of microtubules is skeletal, maintaining a network of membrane 'anchorage' points which may restrict or release surface receptors, depending on both extra- and intracellular stimuli.

Plasma membranes of microbial and plant cells are, of course, commonly enveloped by well-defined cell walls, and it is possible that

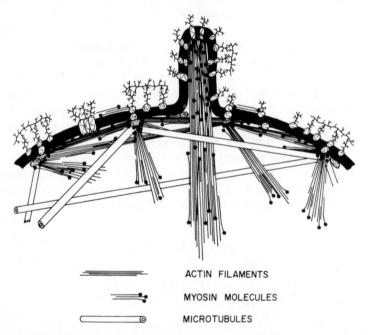

ACTIN FILAMENTS

MYOSIN MOLECULES

MICROTUBULES

Figure 6.16. Hypothetical cytoskeletal network by means of which the lateral movement of cell surface receptors might be controlled. Contractile (microfilaments, shown as actomyosin complexes) and skeletal (microtubules) elements are linked to each other and to plasma membrane components at specific anchorage points. Surface receptors are also seen to be linked via extracellular peripheral proteins in a network that could extend out into the extracellular space (see p. 161): from G. L. Nicolson, *Biochim. Biophys. Acta* (1976), **457**, 57–108.

external elements of this type could also influence the mobility of integral membrane components. Figure 6.16, based upon animal cells, shows interconnection of surface receptors via external peripheral membrane proteins, and it is well known that the region surrounding animal cells includes a carbohydrate-containing matrix. This is made up of molecules such as fibronectin (page 161), collagen and glycosaminoglycans (acidic long-chain polysaccharides associated with protein in proteoglycans) which are not normally classed as cell-membrane components but which may well interact with integral membrane proteins, and through them with the cytoskeleton (see page 161). Transmembrane interactions of this type are conceptually attractive in potentially mediating information transfer between the extracellular environment and the inside of the cell, particularly as the cytoskeleton is now believed to be involved not only in the lateral disposition of membrane components but also in the fundamental processes of cell adhesion and movement.

Membrane asymmetry

The concept of information transfer between surface receptor proteins and cytoplasmic elements implies that the membrane is basically asymmetric. We have already seen that the proteins of the erythrocyte membrane display absolute asymmetry. Thus most of the erythrocyte membrane proteins are associated exclusively with the cytoplasmic face of the membrane, whereas the two major integral proteins, glycophorin A and band 3 polypeptide, span the membrane, displaying different groupings at its inner and outer surfaces. This asymmetry has been found to extend to all membrane proteins so far studied. Once established (see the next section), it might be expected that the asymmetry of membrane proteins will be maintained, as inversion of orientation would require that many charged and polar groups [e.g. most extracytoplasmic parts of membrane proteins carry carbohydrate (page 93)] be forced to pass through the hydrophobic core of the lipid bilayer.

It has only recently become clear that lipids also are asymmetrically distributed within the membrane bilayer. Again the erythrocyte membrane serves as a convenient experimental model in view of its large-scale availability in pure form, and the relative ease with which its inner and outer surfaces can be selectively modified (page 95). Experiments using amine-labelling reagents and also phospholipases have established that the aminophospholipids, phosphatidylethanolamine and phosphatidylserine, are mainly in the cytoplasmic monolayer of the bilayer, whereas the choline-containing phospholipids, phosphatidylcholine and sphingo-myelin are largely externally-localized (figure 6.17a). These results have been confirmed by the use of *phospholipid exchange proteins* which enable phospholipid distribution to be studied without covalent modification of the membrane. Phospholipid exchange proteins occur in the cytoplasm of most eukaryotic cells and act as soluble carrier proteins, catalysing the transfer of specific phospholipids between membranes. The exchange proteins cannot cross membranes of sealed vesicles, and it is accordingly possible to determine the fraction of a given phospholipid in

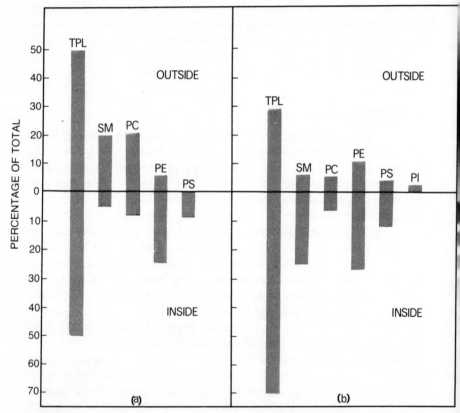

Figure 6.17. Asymmetric distribution of phospholipids in membranes of human erythrocytes (*a*) and of influenza virus grown in bovine kidney cells (*b*). TLP = total phospholipid; PC = phosphatidylcholine; SM = sphingomyelin; PE = phosphatidylethanolamine; PS = phosphatidylserine; PI = phosphatidylinosotol: from J. E. Rothman and J. Lenard, *Science* (1977), **195**, 743–753.

the outer monolayer of a uniformly radiolabelled membrane by exchange with unlabelled phospholipid vesicles.

Not all biological membranes are as suited as that of the erythrocyte to studies of membrane asymmetry. It is important that the membrane is pure and, as discussed in chapter 4, only a limited number of systems are well characterized in this respect. Enveloped viruses represent one such system. The enveloped virus gains its lipid bilayer by budding out from the plasma membrane of the host cell (page 59) and so contains lipids representative, both in composition and orientation, of the host cell membrane. By growing host cells for several generations in radioactive media, it is possible to label viral lipids uniformly and so to measure the distribution of label in the outer monolayer of the membrane by using both exchange proteins and selective enzyme (e.g. phospholipase) modification. These means have been used to determine the orientation of phospholipids in the membrane of influenza virus grown in bovine kidney cells, when the results shown in figure 6.17b were obtained. As can be seen, the distributions differ markedly from those in the erythrocyte membrane. Not only are individual phospholipids differently located in the viral membrane (e.g. sphingomyelin is largely in the inner monolayer) but the total phospholipid content is also asymmetrically distributed, with most of it in the inner monolayer. The resulting lack of phospholipid in the outer monolayer of the viral membrane in these studies was found to be offset by exceptionally high levels of glycolipid. Studies on the distribution of other lipids in both the erythrocyte and viral membranes have shown cholesterol to be approximately equally divided between inner and outer monolayers, whereas the glycolipid, as is generally the case (page 93), was exclusively confined to the outer surface of the membranes.

Bacterial membranes, particularly the single cell membranes of Gram-positive bacilli, can also be used to study lipid asymmetry. Examination of *Bacillus megaterium* has indicated that of its major phospholipids, phosphatidylglycerol predominates in the outer monolayer, and phosphatidylethanolamine predominates in the inner monolayer of the cell membrane.

It appears that lipid asymmetry is a general property of biological membranes but that the detailed nature of this asymmetry varies with the membrane type. Maintenance of lipid asymmetry depends upon the rate of translation of lipid molecules from one monolayer to the other, so-called 'flip-flop' motion. Although for reasons similar to those mentioned in the case of membrane proteins (page 121), such motion might be expected to be generally thermodynamically unfavourable, striking variations in rates of transmembrane movement have been observed in different membranes. Thus phospholipid vesicles and influenza virus membranes have been studied in experiments designed to detect the reappearance of radioactively-labelled phospholipids in outer membrane monolayers previously depleted of labelled phospholipids by exchange proteins. In neither case could 'flip-flop' be detected, and estimates of the half-time of the process gave values of between 10 and 30 days, depending on the phospholipid. Application of other techniques (see e.g. page 245) has led to similar results for the rates of 'flip-flop' of phospholipids in vesicles and black lipid films, and for cholesterol in vesicles and influenza virus membranes. In contrast, half-times for transmembrane movement of phosphatidylcholine in erythrocytes have been estimated at only a few hours, while even shorter half-times (of the order of minutes) have been reported for similar movement of phosphatidylethanolamine in growing bacterial membranes. The rapid transmembrane movement of phospholipid molecules in bacterial membranes suggests the possibility that membranes capable of growth might possess special mechanisms to effect this transfer. Transmembrane integral proteins, for example, might provide hydrophilic channels with which specific phospholipid head groups could interact during their passage through the membrane core. Viral membranes are incapable of growth and, in the absence of such facilitating mechanisms, could be expected to show only the inherent rate of 'flip-flop' of phospholipid bilayers as seen in synthetic vesicles. Demonstration of the existence of specific transmembrane transfer-promoting agents would certainly help to clarify the mechanisms of membrane biogenesis by which lipid bilayers are apparently assembled from precursor molecules on one side of the membrane only (see next section) while at the same time posing further questions concerning the maintenance of lipid asymmetry in membranes containing such agents.

Assembly of biological membranes

There is in most cells a continuous turnover of membrane material and many details of this process are still unclear. Particular problems arise in understanding the ways in which polar membrane components can be transported from a site of synthesis on one side of a basically-hydrophobic lipid bilayer to their final destination on the other side. Membrane glycoproteins, for example, are known to be mainly exposed on the extracytoplasmic faces of plasma and intracellular membranes, whereas the ribosomal synthesizing machinery is clearly in the cytoplasm. Similar problems have long been recognized concerning the biosynthesis of secreted proteins, and it is largely on the basis of evidence concerning secreted proteins that analogous mechanisms for membrane glycoprotein biogenesis in eukaryotic cells have been postulated.

According to modifications of the *signal hypothesis* developed for secreted proteins, synthesis of mammalian membrane glycoproteins begins on free ribosomes in the cytoplasm. The N-terminal section of the nascent peptide carries a hypothetical 'signal' sequence which directs the ribosomes to receptors (probably protein) on the endoplasmic reticulum membrane. As the newly membrane-bound ribosome moves along the mRNA, translation occurs, and the growing peptide is gradually inserted into and through the membrane via a hydrophilic channel associated with the signal receptor. It is suggested that the initial N-terminal hydrophilic peptide section is followed by a hydrophobic sequence (page 92) which, after dissipation of the hydrophilic channel, serves to anchor the integral protein in the membrane. The biosynthesis of secreted proteins differs in that extrusion of the polypeptide continues until it is completely ejected into the intravesicular space. After the hydrophobic section of a membrane protein has been locked into the membrane, synthesis of the C-terminal tail is completed, and the ribosome is released from the membrane and subsequently from the mRNA. At some point following the introduction of the N-terminal peptide into the intravesicular space, the signal sequence may be removed by a specific peptidase ('signalase') and initial sugar residues of the oligosaccharide complexes are probably added (page 140). In the case of membrane integral proteins that are believed to cross the membrane more than once, the N-terminus must somehow re-enter the endoplasmic reticulum membrane—a process that need not affect the intravesicular incorporation of carbohydrate. The membrane glycoprotein now migrates by some means from the rough endoplasmic reticulum to the Golgi apparatus, where carbohydrate incorporation is largely completed. Such migration could take place by vesiculation of special regions of endoplasmic reticulum followed by migration of vesicles to and fusion with the Golgi membranes, or even in some instances by lateral diffusion along the membranes of interconnecting tubular systems. In either case the extracytoplasmic orientation of the membrane glycoprotein is maintained, and both mechanisms could incorporate selection procedures to account for observed differences in membrane composition between different endomembrane systems. Some endoplasmic reticulum membrane proteins, e.g. glucose-6-

phosphatase, are generally absent from Golgi membranes and it is intriguing to speculate that lateral sorting out of proteins within the plane of the endomembranes might be directed by cytoskeletal elements (page 119) prior to dispatch. After completion of their oligosaccharide structures within the Golgi apparatus, membrane glycoproteins are incorporated into vesicles which migrate to, and fuse with, the plasma membrane, so establishing the glycoproteins on its outer face. Secreted glycoproteins, free within the vesicles, would be shed outside the cell at this stage. Many aspects of the overall process (summarized in figure 6.18) are still speculative, but the scheme clearly provides a stimulating basis for further experimentation.

The biosynthesis of membrane components other than extracytoplasmic glycoproteins can, to some extent, be fitted into the above

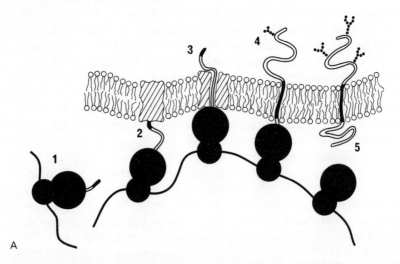

Figure 6.18. Scheme for the biosynthesis of plasma membrane integral glycoproteins. A. Incorporation of the polypeptide into endoplasmic reticulum membrane.
Peptide synthesis begins in the cytoplasm on free ribosomes (1) from which emerge growing polypeptide chains each carrying an N-terminal 'signal sequence' of hydrophobic amino acids. The 'signal sequence' is recognized by an endoplasmic reticulum membrane receptor, probably a protein, which attracts the polypeptide (together with attached ribosome) (2) and facilitates its passage through the lipid bilayer (3). The initially-inserted peptide sequence is followed by a hydrophobic section which, after dissipation of the membrane receptor/channel, anchors the growing polypeptide in the bilayer (4). Following insertion into the intravesicular space, the 'signal sequence' is cleaved off and initial carbohydrate residues (•) are added from lipid precursors (see p. 140) (4). Finally the C-terminal section of the polypeptide is completed and the ribosome is released from the membrane (5) and subsequently from the mRNA.

scheme. Integral proteins associated with the cytoplasmic faces of cell membranes may be synthesized on free ribosomes, whence they could simply diffuse to the appropriate membrane and spontaneously insert a hydrophobic tail sequence. Some recognition mechanism would be required to ensure selection of the correct target membrane, and it is possible that a 'signal sequence' might function in this system also. Alternatively, a hydrophobic insertion sequence on the diffusing protein might become 'activated' by a proteolytic enzyme on the target membrane.

The synthesis of lipids of the extracytoplasmic monolayer of cell membranes poses problems similar to those faced in the case of membrane glycoproteins, in that the enzymes and precursors of membrane lipid are found on the cytoplasmic side of the membrane, at least in bacterial membranes. The small size of membrane lipids precludes insertion mechanisms of the type described above for proteins, and we are left to consider 'flip-flop'-type transfers across the membrane. The slow rates of 'flip-flop' measured in model systems seemed to discount a

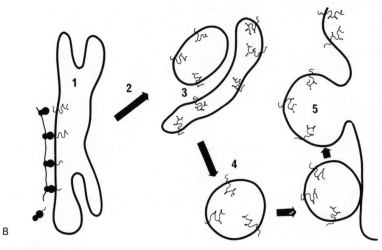

B

Figure 6.18 (continued)

B. Transfer of the polypeptide from the endoplasmic reticulum membrane to the plasma membrane. After completion of the sequence of events outlined in A, membrane-bound partially-glycosylated polypeptide (1) is transferred (2) to the Golgi apparatus where carbohydrate addition is completed (3). The glycoproteins, still membrane-bound, are then packaged in vesicles (4) which merge and fuse with the plasma membrane (5). Throughout the entire process the asymmetric orientation of the glycoprotein with respect to the cytoplasm is maintained.

biosynthetic role for such a process until the recent discoveries of rapid transmembrane movement in bacterial membranes suggested the further possibility that this transfer might be catalysed in certain growing membranes by specific proteins (page 124). In animal cells, lipids are primarily synthesized in the endoplasmic reticulum membrane and it is likely that, as in bacteria, the synthesizing machinery is confined to the cytoplasmic side of the membrane. In this case the problems of post-synthetic lipid distribution are similar to those of bacterial membrane lipids.

Many questions of membrane assembly remain unanswered. No 'flip-flop' promoting agent has yet been identified, and the demonstration of rapid 'flip-flop' in certain membranes has raised the question of how the well-established lipid asymmetry in such membranes is maintained. It may be that asymmetry in these cases reflects a state of dynamic equilibrium in which association of specific phospholipids with one or the other monolayer is thermodynamically preferred. Other problems include the mechanisms of transfer of membrane components, both protein and lipid, between intracellular membranes. The route of a membrane glycoprotein, traced above, from endoplasmic reticulum via the Golgi apparatus to the plasma membrane (figure 6.18) is believed to represent a common progression of membrane components. The details of such *membrane flow* processes are far from clear. Although migrating vesicles are almost certainly involved, especially between the Golgi apparatus and the plasma membrane, it is not established whether individual vesicles operate a back-and-forth *shuttle* system carrying certain transferable components, or whether each vesicle loses its identity by incorporation into the target membrane. The question of selection also remains unsolved. Selection of membrane components for transfer takes place not only in the endoplasmic reticulum but also in the Golgi membranes, and the means by which this process is organized are obscure. All these and further aspects of the membrane assembly process are currently being actively investigated in a number of laboratories, and it is likely that the picture will be further clarified in the next few years.

SUMMARY

1. The *Davson-Danielli-Robertson model* of membrane structure proposed a unit membrane based on a lipid bilayer sandwiched between two layers of protein.

2. There is good evidence, particularly from model systems, for the presence of a bilayer in natural membranes. However, whereas some membrane protein is undoubtedly arranged external to the lipid bilayer as described by Davson, Danielli and Robertson, a number of 'integral' proteins are immersed in and occasionally cross the lipid bilayer as proposed in the *mosaic model*.

3. The lateral mobility of integral proteins within the plane of the lipid bilayer was stressed by the *fluid mosaic model* of Singer and Nicolson. Whereas lateral mobility of both lipids and proteins has been amply demonstrated by physical techniques, it is now clear that such motion, particularly of proteins, is not always random and can be subject to control by both intra- and extra-cellular peripheral protein components.

4. Both proteins and lipids are asymmetrically distributed between the two halves of natural lipid bilayer membranes. 'Flip-flop' motion of lipids from one monolayer to the other is very slow in model vesicles and viral membranes, but can be rapid in growing bacterial membranes where a specific protein-mediated mechanism could operate.

5. Newly-synthesized protein and lipid membrane components of animal cells are initially inserted into the endoplasmic reticulum membrane, whence they are transported, largely in the form of bilayer vesicles, to the Golgi apparatus and hence to the plasma membrane in a general system of *membrane flow*.

CHAPTER SEVEN

GLYCOPROTEINS, GLYCOLIPIDS AND CELLULAR RECOGNITION

ANIMAL CELL PLASMA MEMBRANES CONTAIN ASYMMETRICALLY-DISTRIBUTED glycoproteins and glycolipids which extend their carbohydrate-bearing portions directly into the extracellular environment, and there is currently a great deal of interest in the possible involvement of such molecules in the many aspects of cell recognition. The cell membranes of bacteria, fungi and higher plants are, in general, in contact with a complex carbohydrate-rich cell wall which complicates characterization of sugar-containing plasma membrane components. The cell wall, moreover, shields the underlying membrane from direct interaction with the cell surroundings, and the present chapter will accordingly be mainly concerned with the structure and function of the glycolipids and glycoproteins of animal cell plasma membranes.

Evidence for the presence of carbohydrate on the surface of cellular membranes

It is worth while to outline briefly some of the methods that have been employed to determine the localization of membrane-associated sugar residues. Cytochemical techniques have been widely used in this respect. The periodic acid-Schiff (PAS) reagent (page 94) gives a purple stain, readily detectable in the light microscope, which is fairly specific for carbohydrate, although prolonged periodic acid oxidation can elicit a false positive PAS reaction from hydroxyamino acids. Periodic acid oxidation is also used to visualize carbohydrate in electron microscopy, when the dialdehyde oxidation product (page 94) is allowed to react with silver methenamine to give an electron-dense silver precipitate. Membrane carbohydrate commonly contains sialic acid and occasionally sulphate anions. A number of microscopic procedures have accordingly been developed which depend upon the presence of such acid groupings. In the Hale stain, colloidal iron is attracted to the acidic groups, where it is allowed to react with potassium hexacyanoferrate (II) giving Prussian blue, detectable by both light and electron microscopy. Alternatively, the bound colloidal iron particles can be directly visualized in the electron microscope without further derivatization. Colloidal thorium has also been used in this way. Ruthenium red has the proposed structural formula $[(NH_3)_5Ru\text{-}O\text{-}Ru(NH_3)_4\text{-}O\text{-}Ru(NH_3)_5]^{6+}$ and undergoes a complex reaction with anionic surface material, including acidic carbohydrate. It has been widely used both in light and, in combination with osmium tetroxide, in electron microscopy (figure 7.1). This use of cationic reagents in carbohydrate detection carries the danger of their inherent lack of specificity and the use of control experiments (e.g. using the enzyme neuraminidase to remove sialic acid) is usually advisable.

More specific techniques for the detection of carbohydrate have resulted from the use of *lectins*. Proteins that can agglutinate erythrocytes and other cell types have been extracted from a wide range of plants, particularly from the seeds of legumes, and to a lesser extent from other sources such as snails, crabs, fish and micro-organisms. Agglutinins from all sources are commonly referred to as *lectins*, and the term will be used in that sense here. Lectins carry multiple binding sites for sugar residues, and bring about agglutination by simultaneously binding to cell-surface carbohydrate on a number of cells, so cross-linking them. Such binding is usually quite specific, each lectin binding only to certain carbohydrate

Figure 7.1. Electron micrograph of a section of human erythrocyte membrane stained with ruthenium red-osmium tetroxide. The very densely stained layer is believed to result from the carbohydrate components of membrane glycoproteins: from G. M. W. Cook and R. W. Stoddart, *Surface Carbohydrate of the Eukaryotic Cell* (1973), London, Academic Press, p. 74, by courtesy of Dr. G. M. W. Cook, Strangeways Research Laboratory, Cambridge.

receptors. Thus concanavalin A, from Jack beans, is specific for α-D-glucopyranosyl and α-D-mannopyranosyl residues, whereas wheat germ agglutinin will only bind to N-acetylglucosamine units. Lectin binding to cells (and hence agglutination) can be inhibited by addition of the relevant sugars (e.g. glucose or mannose in the case of concanavalin A) which compete with cell-bound residues for the lectin and so displace it from the cell. Inhibition of agglutination by a given sugar implies the presence of that sugar on the cell surface and *hapten inhibition* studies of this type were used by Watkins and Morgan in the 1950s to define the carbohydrate determinants of ABO blood group activity on red blood cells (page 144). The early work of Watkins and Morgan was, in fact, among the first pieces of evidence for the presence of sugars on cell surfaces.

Binding of lectins to cell membranes can be determined directly by labelling the lectin in various ways. Thus fluorescein-labelled lectins can be incubated with cells, and the bound lectins visualized on the cell surface by fluorescence microscopy. Alternatively, quantitative assays of cell-lectin binding can be effected using radioactively-labelled lectins. The binding of lectins to cell surfaces has also been detected by electron microscopy by coupling electron-dense material to the lectin both before and after attachment to the cell; for example, cell-bound lectins have been visualized by making use of the fact that lectins have more than one binding site. Horseradish peroxidase, a glycoprotein, is allowed to bind to the unoccupied sites of lectins already attached to a cell surface. The catalytic activity of the lectin-bound peroxidase is then utilized to produce an electron-dense product by reaction with diaminobenzidine. A similar but rather more direct procedure involves attachment of haemocyanin via its carbohydrate residues to membrane-bound lectin, and stabilization of the complex with glutaraldehyde. Haemocyanin is a large molecule of distinctive shape which can be clearly seen in shadow-cast replicas (figure 7.2). Electron-dense markers can also be covalently linked to lectins before attachment of the conjugate to the cell membrane. Haemocyanin, ferritin (an iron-storage protein, figure 6.12), and gold granules have all been used in this way to produce maps of lectin binding sites on the cell surface. As in lectin-induced agglutination, the specificity of labelled-lectin binding can always be checked by control experiments in which the observed interaction is inhibited by added carbohydrates acting as haptenic inhibitors (see e.g. figure 7.2). In this way evidence for the presence of individual sugar receptors on cellular membranes can be obtained.

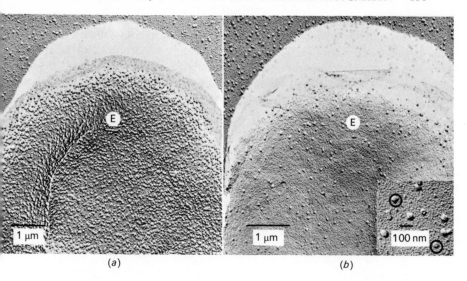

(a)   (b)

Figure 7.2. (*a*) Rat erythrocyte (E) treated with concanavalin A followed by haemocyanin. The haemocyanin molecules are distributed evenly over the surface of the cell.
(*b*) Rat erythrocyte (E) treated with concanavalin A plus methyl α-D-glucoside, then haemocyanin. This is the control to figure (*a*). Methyl α-D-glucoside inhibits concanavalin A binding so that very few haemocyanin molecules are seen on the cell. The amount of haemocyanin seen in the background on the glass is unchanged. The inset is a higher magnification showing that the haemocyanin molecules can be clearly recognized. As they are cylindrical, the molecules seen from the side appear rectangular and seen end on circular. Representatives of these two views have been circled: from S. B. Smith and J. P. Revel, *Development Biology* (1972), **27**, 434–441, by courtesy of Dr. S. Smith Brown, Division of Biology, California Institute of Technology.

The presence of sialic acid on the surface of human erythrocytes was demonstrated by the use of *cell electrophoresis* in the early 1960s. In this technique, cells are suspended in an electrolyte solution and subjected to an electric field, when the mobility of the cells reflects their surface potential. Under these conditions, human erythrocytes migrate toward the anode, indicating a net negative charge on the cell surface. The variation of electrophoretic mobility with pH is shown in the profile of figure 7.3 (upper curve), the steep part of which reflects the progressive ionization of an acidic group with pK of approximately 2·5. This group has been identified as sialic acid (pK 2·6) by treatment of erythrocytes with neuraminidase when their electrophoretic mobility falls to that of the lower curve (figure 7.3). The electrophoretic mobility of erythrocytes is also reduced by treatment with the proteolytic enzyme trypsin, which releases sialic-acid-containing glycopeptides into the medium, indicating that sialoglycoprotein (page 95) makes a major contribution to membrane surface charge. The negative mobility (figure 7.3, lower curve) remaining after neuraminidase treatment can be attributed to the presence of ionized glycolipid sialic acids (not always susceptible to neuraminidase) or to acidic groupings of higher pK, such as the car-

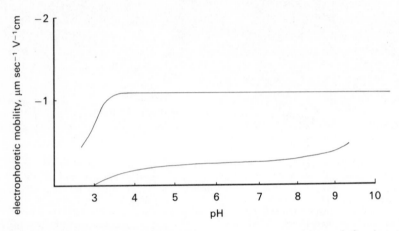

Figure 7.3. pH-electrophoretic mobility relationships for human erythrocytes before (upper curve) and after (lower curve) neuraminidase treatment: from G. M. W. Cook *et al.*, *Nature* (1961), **191**, 44–47.

boxylic side-chains of protein glutamic acid or aspartic acid residues. pH-electrophoretic mobility curves of other cell types are commonly more complex than that of the erythrocyte, and the profiles obtained using lymphocytes, blood platelets, cultured mammalian cells and milk fat globules (page 49) show additional features (e.g. mobility to the cathode at low pH) resulting from more significant contributions to the overall surface charge from protein and other groups.

As a result of the application of the various techniques described above, it has become clear that most cellular membranes carry carbohydrate complexes which are located almost exclusively on the extracytoplasmic faces of both plasma and intracellular membranes. This carbohydrate results from the presence of glycolipid and glycoprotein molecules, and the isolation, structures and biosynthesis of these compounds will be discussed in the next sections.

Extraction and purification of membrane glycolipids and glycoproteins

Lipids are commonly extracted from tissues by chloroform-methanol mixtures, or by ethanol, and are partitioned in a biphasic chloroform-methanol-water (or dilute salt solution) system. Following this 'Folch' partition, most of the lipids, including the smaller neutral glycolipids, are contained in the lower chloroform-methanol layer, whereas gangliosides and larger neutral glycosphingolipids remain in the upper water-methanol layer. Complex blood group lipids (page 147) have been separated in the water phase of butanol-water mixtures also. Gangliosides can be separated from uncharged glycolipids on columns of DEAE-cellulose, while further purification of all glycolipids is usually effected by column and thin-layer chromatography on silicic acid.

Membrane glycoproteins are often integral rather than peripheral proteins, and can be difficult to obtain in a soluble form free from solubilizing agent without denaturation or aggregation (page 90). Some glycoproteins, however, can be obtained in water-soluble form by means of various extraction procedures, exemplified by those used to prepare the glycophorins, the major sialoglycoproteins of the human erythrocyte membrane (page 95). Glycophorins have been extracted by a range of organic solvent-water mixtures, including butanol-water, pentanol-water, phenol-water and chloroform-methanol-water systems, when the glycoproteins are found in the aqueous phase. Lithium diiodosalicylate (LIS), a combination of detergent and chaotropic agent (page 90), has also been employed in the preparation of glycophorins, as has the organic solvent pyridine. When, like the glycophorins, a membrane glycoprotein can be extracted into true aqueous solution, then further purification is effected by standard techniques of protein chemistry. Many integral glycoproteins can only be extracted and maintained in solution in the presence of detergents (e.g. Triton, sodium dodecyl sulphate), organic solvents or chaotropic agents, in which case procedures for their further fractionation in the continued presence of the solubilizing agents have had to be developed (chapter 5). Affinity chromatography using immobilized lectins (figure 5.39) such as concanavalin A and lentil lectin (both specific for glucose and mannose) or wheat germ agglutinin (specific for N-acetylglucosamine) has proved to be a particularly powerful technique for the purification of membrane glycoproteins generally.

'Purification' of a membrane glycoprotein seldom implies isolation of a single chemically-defined molecular type. Minor amounts of similar glycoproteins and even tightly-bound glycolipid often accompany a membrane glycoprotein through several fractionation steps, and structural studies must be carried out with this in mind. Present evidence indicates that even a fraction that is homogeneous in its polypeptide chains is likely to be heterogeneous with respect to its carbohydrate content. *Microheterogeneity* of this kind has been demonstrated in many soluble glycoproteins which contain a range of oligosaccharide side-chains, representing minor modifications and various stages of 'completion' of a theoretical 'complete' unit. Such 'incomplete' carbohydrate structures could arise either from relative inefficiency of glycosyl transferases during biosynthesis or from removal of terminal sugar units by glycosidases after biosynthesis of the 'complete' carbohydrate complex. In view of the role played by carbohydrate in a number of recognition phenomena to be discussed in this chapter, it is conceivable that carbohydrate microheterogeneity might generally reflect functional differences between individual molecules in a pool of glycoproteins although there is currently little evidence for this.

Structures of membrane glycolipids and glycoproteins

Structural analysis of purified glycolipids and glycoproteins is based on standard procedures of lipid, protein and carbohydrate chemistry. Examination of the carbohydrate fraction of glycoproteins requires separation of the oligosaccharides from the bulk of the protein. This is commonly achieved by extensive proteolysis of glycoproteins, made soluble as described above, or of glycopeptides proteolytically-cleaved direct from the membrane surface. Glycopeptides obtained by these means are water-soluble and can be further purified by gel-filtration, ion-exchange chromatography, etc. Purification is simplified if the carbohydrate can be obtained free from peptide, and certain oligosaccharide-peptide linkages can be readily cleaved by mild alkali (see page 137). In recent years methods have also been developed for both chemical and enzymic (endoglycosidase or glycopeptidase) release of oligosaccharides from 'alkali-stable' glycopeptide linkages (page 137). The detailed carbohydrate structures of purified glycolipids, glycopeptides or oligosaccharides are examined by conventional techniques such as periodate oxidation, methylation and glycosidase cleavage procedures, coupled with gas chromatographic and, more recently, mass spectrometric analysis.

The basis of the structural types of membrane glycolipids has been outlined in chapter 5. Although detailed analyses of the oligosaccharide

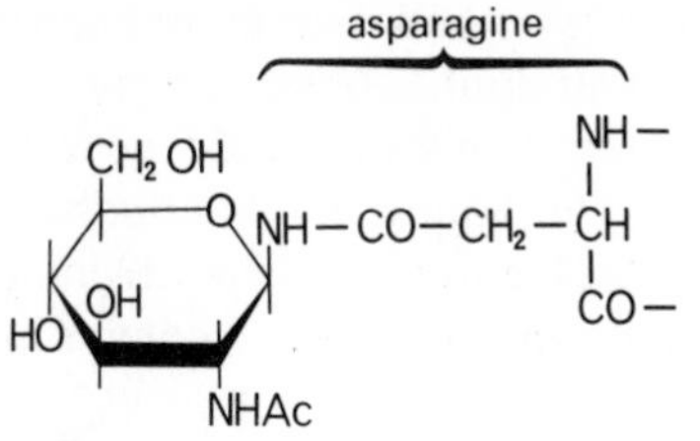

Figure 7.4. Alkali-stable glycopeptide bond involving an N-glycosidic linkage between asparagine and N-acetylglucosamine.

structures of only a few membrane glycoproteins have been done, a general pattern, very similar to that of secreted glycoproteins, is beginning to emerge. The carbohydrate content is made up of some or all of a group of just six monosaccharide units: D-galactose, D-mannose, L-fucose, N-acetyl-D-glucosamine, N-acetyl-D-galactosamine and sialic acid (acylated neuraminic acids, commonly N-acetylneuraminic acid or N-glycolylneuraminic acid), arranged in oligosaccharide side-chains of varying sizes and glycosidically-linked to a polypeptide backbone. Carbohydrate-peptide linkages are of two major types, both of which may occur together within the same glycoprotein. One linkage involves an N-glycosidic bond from N-acetylglucosamine to the amide nitrogen of asparagine (figure 7.4) and is relatively stable to hydrolysis by strong alkali. The second type of linkage contains an O-glycosidic bond from N-acetylgalactosamine to the hydroxy-group of serine or threonine (figure 7.5) and is cleaved by mild alkaline treatment. Alkaline cleavage involves β-elimination of the glycosidically-linked sugar moieties as shown in figure 7.6, leaving a double-bonded amino-acid residue in the peptide chain. In structural studies, treatment with alkaline borohydride is often preferred to simple alkali, as the released N-acetylgalactosamine

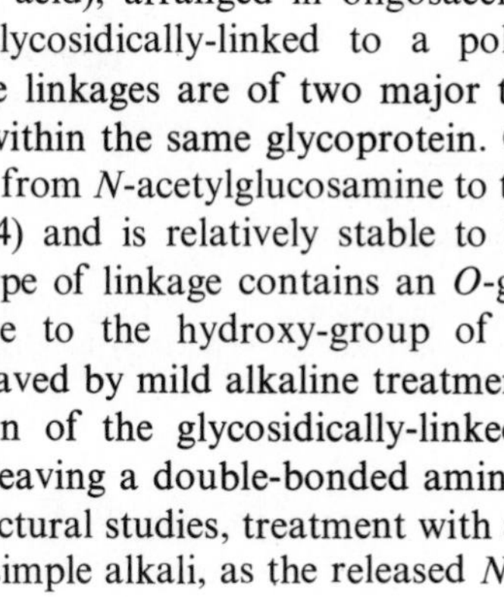

Figure 7.5. Alkali-labile glycopeptide bond involving an O-glycosidic linkage between serine or threonine and N-acetylgalactosamine.

is protected from alkaline degradation by reduction to N-acetylgalactos-aminitol. Moreover, the double-bonded amino acids, 2-aminoacrylic acid (from serine) and 2-aminocrotonic acid (from threonine), are reduced to the more stable alanine and 2-aminobutyric acids respectively.

A further type of O-glycosidic linkage, from galactose to hydroxy-lysine, is found in basement membrane proteins and collagens.

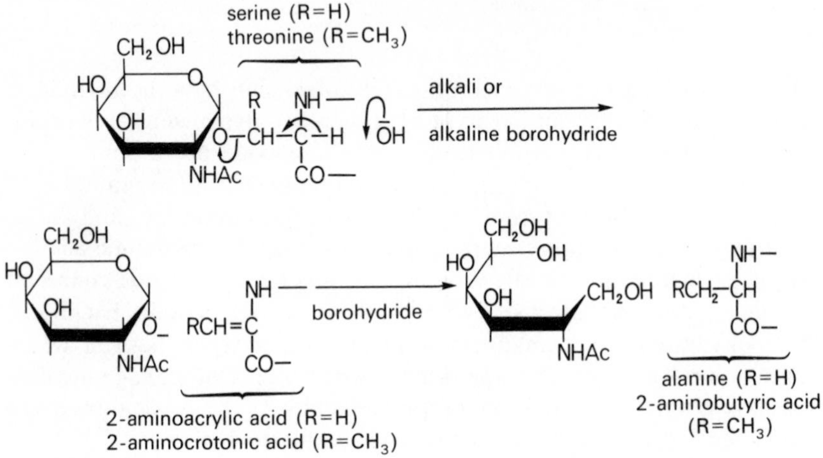

Figure 7.6. Alkali-induced β-elimination of N-acetylgalactosamine from serine or threonine, and subsequent reduction of the cleavage products by borohydride.

A common structural pattern for the N-glycosidically-linked oligo-saccharide complex of secreted glycoproteins is shown in figure 7.7, and the limited number of corresponding structures that have been worked out for membrane glycoproteins seem to represent variations on this general theme (see e.g. figures 7.13 and 7.17). Oligosaccharide complexes that are O-glycosidically-linked to membrane glycoproteins tend to be relatively simple, and the tetrasaccharide shown in figure 7.8 has been detected in membrane glycoproteins from human, horse and sheep erythrocytes (e.g. glycophorin A, page 96), milk fat globules and rat brain. More complex O-glycosidically-lined structures have been reported in the cell membranes of bovine erythrocytes and on a mouse tumour cell line.

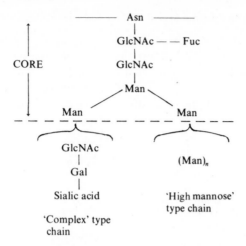

Figure 7.7. Schematic representation of a typical *N*-glycosidically-linked oligosaccharide complex in secreted glycoproteins. The 'core' structure is usually elongated by addition of *N*-acetyllactosamine units which may carry terminal sialic or fucose ('complex' type chains) or by addition of polymannosyl units ('high mannose' type chains). A single glycoprotein may have both types of chain elongation, and up to four chains can branch off from the mannose units of the core. Membrane glycoproteins appear to contain similar structures; adapted from J. Montreuil, *Pure Appl. Chem.* (1975), **42**, 431–477.

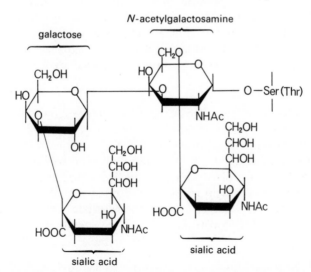

Figure 7.8. Structure of an *O*-glycosidically-linked tetrasaccharide complex found in a number of membrane glycoproteins.

Biosynthesis of membrane glycoproteins and glycolipids

The mechanisms of assembly of eukaryotic glycoprotein oligosaccharide complexes have, over the last few years, become the object of considerable research activity—activity that was largely stimulated by the discovery of the involvement of *dolichols*, a family of long-chain isoprenyl alcohols (figure 7.9) in the biosynthetic process. Dolichol is now believed

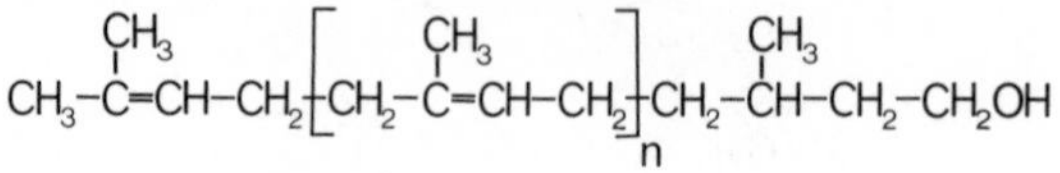

$$\underset{CH_3}{CH_3-\overset{CH_3}{\overset{|}{C}}=CH-CH_2}\left[CH_2-\overset{CH_3}{\overset{|}{C}}=CH-CH_2\right]_n CH_2-\overset{CH_3}{\overset{|}{C}H}-CH_2-CH_2OH$$

Figure 7.9. General structure of the dolichols, $n = 14 \rightarrow 20$. The term 'dolichol' is used generally to describe mixtures of homologues which, in the case of animal biosynthetic intermediates, usually have $n = 17 \rightarrow 20$.

to act as an intermediate carrier for sugar residues in the assembly of N-glycosidically-linked carbohydrate side-chains, as in the scheme for biosynthesis of mammalian and avian glycoproteins outlined in figure 7.10.

In this scheme two N-acetylglucosamine units (the first as its α-phosphate) and the first β-linked mannose residue are transferred sequentially from sugar nucleotide precursors to a dolichol phosphate acceptor. Eight further mannose units are individually transferred from GDP-mannose to dolichol monophosphate, and thence to the growing oligosaccharide chain on its dolichol pyrophosphate carrier. Two or three glucose residues are added in a terminal sequence by a similar mechanism to complete the common lipid intermediate

$$\text{Dol—P—P} \overset{\alpha}{-} \text{GlcNAc} \overset{\beta}{-} \text{GlcNAc} \overset{\beta}{-} \text{Man—(}\alpha \text{ Man)}_8 \text{—(}\alpha \text{ Glc)}_{2-3}$$

(figure 7.10) for which the oligosaccharide structure shown in figure 7.11 has been proposed. The oligosaccharide moiety is now cleaved from its dolichol pyrophosphate carrier and linked N-glycosidically to asparagine on a polypeptide precursor. This transfer of oligosaccharide from lipid to protein probably takes place within the endoplasmic reticulum following insertion of the newly synthesized polypeptide into the intravesicular space (page 125), but whether the lipid intermediate is assembled on the cytoplasmic or extracytoplasmic side of the membrane remains un-

Figure 7.10. Possible general pathway for the biosynthesis of N-glycosidically-linked oligosaccharides of secreted and membrane-bound mammalian and avian glycoproteins. Dol = dolichol; P = phosphate.

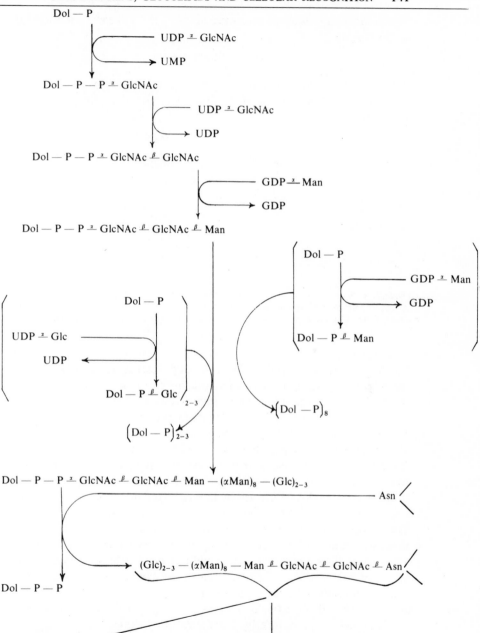

determined. The sequence of amino acids at the linkage region always has the form:

—asparagine—X—serine (threonine)

where X can be almost any amino acid, and it is generally assumed that this sequence represents a recognition site for the oligosaccharide-transferring enzyme.

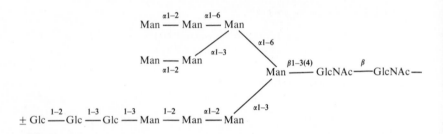

Figure 7.11. Structure proposed for the oligosaccharide moiety transferred from dolichol pyrophosphate to polypeptide in the biosynthesis of G protein of vesicular stomatitis virus-infected Chinese hamster ovary cells: from E. Li, I. Tabas and S. Kornfeld, *J. Biol. Chem.* (1978), **253**, 7762–7770.

After its transfer to polypeptide, the oligosaccharide must be further modified by cleavage of glucose and mannose residues in order to allow synthesis of 'complex'-type chains (figure 7.7), and at least the glucose units must be removed en route to 'high-mannose' chains also. Work on the biosynthesis of complex-type oligosaccharides of vesicular stomatitis virus G protein in Chinese hamster ovary cells has indicated a possible sequence of modification. It is proposed that processing of the protein-bound oligosaccharide shown in figure 7.11 is initiated by rapid removal of the glucose residues, followed by the 1-2 linked mannose units, to leave a residual complex containing five mannoses. An *N*-acetylglucosamine residue is then transferred to one of the α 1-3 mannose units (that shown on the bottom line of figure 7.11), and the two remaining terminal mannose residues are cleaved before addition of a further *N*-acetylglucos-amine and elaboration of the 'complex-type' chains. Most of the processing probably occurs in the Golgi apparatus where it is believed that the terminal *N*-acetylglucosamine, galactose, sialic acid and fucose are all added by direct transfer from sugar nucleotides.

It is of interest that, provided the glycosidic link is broken, every monosaccharide transfer shown in figure 7.10 involves inversion of anomeric configuration. So because UDP-N-acetylglucosamine and GDP-mannose are both α-anomers, incorporation of a β-linked sugar unit into the growing oligosaccharide chain can occur directly from the sugar nucleotide, whereas the α-linked mannose residues are added via dolichol monophosphate β-mannose with double inversion of configuration. The first N-acetylglucosamine unit is added to dolichol phosphate as its α-phosphate, and anomeric inversion occurs only in the final transfer to the peptide chain.

Evidence for a biosynthetic scheme of the type described above has been found in mammalian, avian and yeast cells and, although details may vary, it is currently believed that the scheme could represent a mechanism for the biosynthesis of both membrane-bound and secreted glycoproteins in eukaryotic cells generally.

The assembly of O-glycosidically-linked side-chains (page 137) of mammalian and avian glycoproteins is probably wholly effected by glycosyl transferases in the Golgi apparatus without the involvement of lipid carriers. No obvious recognition sequence of amino acids corresponding to that of the N-glycosidically-linked side-chains has been detected in the region of the serine (threonine)—N-acetylgalactosamine linkage (figure 7.5). The alkali-labile mannans of yeast envelope glycoproteins contain mannose-serine (threonine) O-glycosidic linkages that, in apparent contrast to mammalian and avian systems, are formed by transfer of mannose from dolichol monophosphate mannose. Further mannosylation probably occurs by direct stepwise addition of mannosyl units from GDP mannose.

Biosynthesis of bacterial peptidoglycan, lipopolysaccharide and other polysaccharides has been known for some time to involve polyprenol-linked sugars. In these cases the lipid carrier is the phosphomonoester of undecaprenol, a C_{55} polyprenol which differs from the longer chain dolichols in that its α-isoprene unit is not saturated (cf. figure 7.9).

The glycolipids of mammalian cell membranes are largely glycosphingolipids which are believed to be assembled by sequential transfer of monosaccharide residues from sugar-nucleotide precursors to ceramide or sphingosine (itself biosynthesized by condensation of a fatty aldehyde or fatty acyl CoA with pyridoxal phosphate-bound serine). It is likely that the oligosaccharide complex is built up by the action of glycosyl transferase enzymes in the Golgi apparatus in much the same way as in the latter stages of N-glycosidically-linked glycoprotein biosynthesis (page 142).

Functions of cell membrane glycoproteins and glycolipids

In view of the presence of complex glycoprotein and glycolipid molecules at the outer surface of animal cells, it might be expected that these molecules are involved in the interaction of the cells with their surroundings. The oligosaccharide complexes of these molecules are composed of six basic monosaccharide types capable of being combined and disposed in infinite variations and it is an attractive concept that carbohydrate patterns might play a major role in cell surface recognition phenomena. In a number of cases this has been found to be so, while in others the contribution of carbohydrate to the demonstrated specificity of certain membrane glycoproteins remains unclear. Examples of both of these situations will be discussed in following sections.

Blood group specific antigens

Since the discovery of the classic ABO system by Landsteiner in 1900, blood group characters have been shown to be inherited according to Mendelian laws and many independent blood group systems have been described. The chemical nature of the antigens determining ABO and the related Lewis (Le^a and Le^b) specificities is now established, and some progress has been made toward clarification of the structural basis of the MNSs blood group system. Little is known about the molecular nature of most other blood group antigens, however, including those of the clinically important Rhesus system.

Classification of the *ABO blood group system* is based upon the occurrence of antigens on human erythrocytes and the relationship of these antigens to the presence of specific antibodies in the blood serum is shown in Table 7.1.

Specific A, B and H antigens are found not only on erythrocytes but also on epithelial and endothelial cells, and as soluble components of

Table 7.1. Antigen and antibody characteristics of the ABO blood group system.

Blood Group	Antigen on red cell	Antibody in serum
A	A	anti-B
B	B	anti-A
AB	A and B	—
O	—*	anti-A and anti-B

* Red cells of group O carry an H antigen against which antibodies are present in some non-human species.

secretions. The antigenic determinants on the surfaces of erythrocytes and other cells are carried by both glycolipids and glycoproteins, whereas those of soluble blood group substances are glycoprotein in nature. Purified membrane glycoconjugates are available on only a small scale and techniques for their isolation and structural analysis have only been fully developed relatively recently. Because of these difficulties, the chemical basis of ABO and Lewis specificities, established in the 1950s, were worked out by using blood group-specific glycoproteins of ovarian cyst fluid, which are available in quantity. The blood group activities of the soluble glycoproteins were assessed by their ability to inhibit agglutination of red blood cells by specific antisera, and by direct precipitation by antisera. In this way it was found that ABO specificity of the soluble glycoproteins was carried by monosaccharide units. The role of membrane-bound monosaccharides in determining the blood group specificity of the red cells themselves was confirmed by hapten inhibition techniques using lectins and antisera (page 132).

Inheritance of ABO and Lewis characteristics is controlled by the action of genes at four independent loci: ABO, Lele, Hh and Sese. It is believed that the A, B, Le and H genes produce glycosyl transferases that attach the relevant immunodominant sugar to a specific precursor as shown in figure 7.12. The Le gene and H gene each produce a fucosyl transferase that attaches a single fucose residue to a specific sugar of the precursor structure so conferring respectively Le^a and H activities. Expression of both the Le and H genes causes addition of two fucose residues producing the Le^b structure. Soluble blood group active glycoproteins contain both type I and type II chain precursors (figure 7.12) but the 4-fucosyl transferase produced by the Le gene clearly cannot act on type II chains, which are accordingly not associated with Lewis activity. The A and B genes produce N-acetylgalactosaminyl transferase and galactosyl transferase respectively, which can only act on the H structure. If the H gene is absent, neither A nor B structures can occur, even though the relevant genes may be present. Provided that the H gene is present, however, the A gene gives rise to the A antigen and the B gene to the B antigen. When both A and B genes are present together, both A and B structures arise, whereas, when both are absent, the H structure remains unaltered and the blood group is termed O.

These relationships have been worked out using soluble glycoproteins that carry their blood-group determinant sugar groupings at the non-reducing ends of complex O-glycosidically-linked oligosaccharide side chains. In fact the appearance of soluble blood group substances with

A, B or H specificity is dependent on the presence of the Se gene. About 80% of the population (so called *secretors*) possess the Se gene which controls the expression of the H gene in secretions. In 'non-secretors' the Se gene is absent, and products of the H-gene are not seen in secretions which consequently (see figure 7.12) do not show A, B, H or Le^b blood

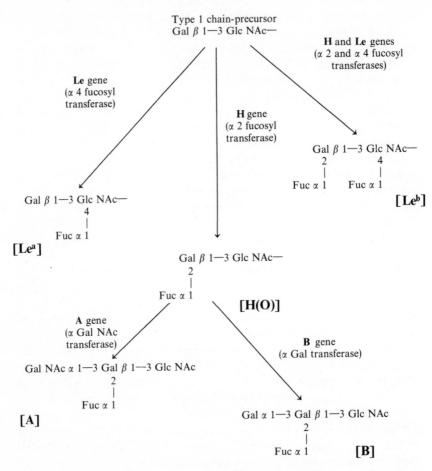

Figure 7.12. Biochemical pathways for the conversion of type I (Gal β 1–3 GlcNAc—) and type II (Gal β 1–4 GlcNAc—) precursors into Le^a, Le^b, A, B and H-active structures. Each gene specifies a glycosyltransferase (given in parentheses) which adds the immunodominant sugar: adapted from W. M. Watkins in *Glycoproteins. Their Composition, Structure and Function*, Part B, Amsterdam, Elsevier (1972), pp. 830–891.

group activity. Non-secretors may have secretions with Le^a activity and can show A, B or H antigens on the erythrocyte surface.

Improvements in analytical techniques in the late 1960s and early 1970s have led to the elucidation of the structures of glycolipid ABO and Lewis determinants on the erythrocyte membrane itself. A, B and H-active glycosphingolipids based on the neolacto-series (Table 5.3) have been characterized. The A, B, and H oligosaccharide determinants shown in figure 7.12 (type II chains only) are carried on the non-reducing ends of oligosaccharide chains of varying complexity, the simplest of which is the Gal β 1-4 GlcNAc terminus of neolactotetraose (Table 5.3, p. 79) itself. Le^a- and Le^b-active glycolipids [which cannot have type II chains (page 145)] are derived in a similar manner from lactotetraose (Table 5.3). More recently, ABH-active glycopeptides have been isolated from erythrocyte membranes and shown to contain complex oligosaccharides remarkably similar to those of the corresponding glycolipids.

In 1927, Landsteiner and Levine described a second blood-group system called MN in which two alleles, M and N, control the occurrence of three genotypes MM, MN and NN. In defining the system Landsteiner and Levine used antisera from rabbits injected with human red blood cells and, as the corresponding antibodies occur only infrequently in man, the MN antigens do not cause major problems in blood transfusion. In 1947 an additional two-allele antigenic system Ss was found which is closely linked with MN, and in recent years the nature of the antigens determined by the *MNSs system* has been intensively studied.

M and N antigenic activity is associated with glycophorin A, the major sialoglycoprotein of the human erythrocyte membrane. Glycophorin A (figure 5.41) contains sixteen oligosaccharide side-chains, all of which are attached to the *N*-terminal polypeptide chain and extend into the extracellular space. Fifteen of these side chains are *O*-glycosidically-linked units and are generally assumed to be either similar or identical to the tetrasaccharide shown in figure 7.8. The sixteenth side chain is an *N*-glycosidically-linked oligosaccharide complex for which the partial structures shown in figure 7.13 have been suggested. It is worth pointing out that both of these structures were proposed at an early stage in a fast-moving research area and that in view of their deviation from what is now believed to be a common glycoprotein 'core' structure (page 139), a re-examination might prove fruitful.

The structural basis for MN antigenic specificity has been hotly debated. Anti-M or anti-N activity is shown by sialoglycopeptides cleaved from the surfaces of the corresponding erythrocytes, but is destroyed by subsequent treatment of the fragments with neuraminidase or alkaline borohydride. This suggests that the MN antigenic determinants include the O-glycosidically-linked oligosaccharides of glycophorin A, and weak but specific M or N activities have been reported for oligosaccharides derived from the corresponding erythrocytes. However, while the alkali-labile carbohydrate, and especially the sialic acid units of

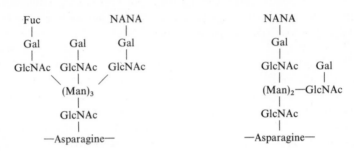

Figure 7.13. Partial structures of N-glycosidically-linked oligosaccharide side chains from human erythrocyte membrane glycoproteins proposed by: (*a*) D. B. Thomas and R. J. Winzler, *Biochem. J.* (1971), **124**, 55–59; (*b*) R. Kornfeld and S. Kornfeld, *J. Biol. Chem.* (1970), **245**, 2536–2545.

glycophorin, are clearly important in antibody recognition, the primary determinants of M- or N-activity appear to lie in the polypeptide chain. Analysis of glycophorin A from individuals with genotypes MM, MN and NN (glycophorins A^M, A^{MN} and A^N respectively) has shown small but consistent differences in amino-acid sequences. Thus glycophorin A^M has the N-terminal sequence Ser-Ser-Thr-Thr-Gly, whereas glycophorin A^N has Leu-Ser-Thr-Thr-Glu, and glycophorin A^{MN} has both structures. Apart from positions 1 and 5, the remaining 129 amino acids of glycophorin A are believed to be identical in the three genotypes. Moreover, cyanogen bromide cleavage of glycophorins has led to the characterization of two N-terminal glycopeptides which contain just eight amino acids and apparently identical carbohydrate moieties, but which show M- and N-reactivities respectively. These can be attributed to the specific amino-acid sequences shown above.

Serological Ss activity is believed to be associated with glycophorin B (page 97) on the red blood cells. Glycophorin B is not as well characterized as glycophorin A, but it can be separated from the latter by gel filtration, and the sequence of its first 22 amino acids has been shown to be identical to that of glycophorin A^N. Glycophorin B from MM or NN cells will, in fact, inhibit the agglutination of NN erythrocytes, and this explains frequent reports of weak N-activity on MM homozygous cells. The structural basis of Ss specificity on glycophorin B is not yet defined.

Clarification of the involvement of glycophorins A and B in MNSs blood group specificities has been greatly helped by the existence of genetic variants which lack one or the other of these erythrocyte glyco-proteins (cf. page 111), and it is an interesting reflection on the possible functions of the glycoproteins that individuals lacking them show no apparent abnormality whatsoever.

Histocompatibility antigens

Histocompatibility antigens are cell surface antigens that are specific for the individual and are involved in the rejection of transplanted tissue. These antigens can be most conveniently studied by controlled breeding in mice, in which over 30 distinct histocompatibility antigens differing in their ability to elicit rejection have been defined. Antigens giving rise to the strongest graft rejection in mice are coded by the H-2 gene complex on chromosome 17 (figure 7.14), and the major histocompatibility antigens (transplantation antigens) are products of the H-2K and H-2D regions of this complex. They are present on most mammalian cells. Man is clearly less amenable to genetic manipulation, and the human system has been less studied. The human counterpart of the H-2 complex is located on chromosome 6 and is called HLA, meaning H for human, L for leucocyte (the cell type first studied), and A for the first system.

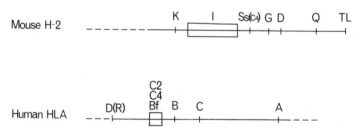

Figure 7.14. A schematic comparison of the genetic maps of the mouse H-2 and human HLA regions. The letters above the lines indicate loci or regions (Ss stands for serum substance and is a gene for the fourth component of complement C4). C2, C4 and Bf are genes coding for the second and fourth factors of the classical complement pathway and for factor B of the alternative pathway respectively: from W. F. Bodmer, *Br. Med. Bull.* (1978), **34**, 212–216.

Three regions of this complex, HLA-A, HLA-B and HLA-C (figure 7.14) code for histocompatibility antigens. [The ABO blood group antigens (page 144) clearly fit the above definition of histocompatibility antigens but have here, as is usual, been considered separately.]

Cell surface antigens produced by the H-2K and H-2D genes are commonly referred to simply as H-2 antigens, and outbred strains of mice will carry four such antigens (one for each of the H-2K and H-2D genes of the parents). H-2 antigens have been extracted from cell membranes by detergent and shown to have the general structure shown in figure 7.15. A transmembrane glycoprotein of approximate molecular weight 45 000 is associated with a small non-glycosylated peptide (molecular weight 12 000) which is identical with the urinary protein β_2-microglobulin. Cleavage of the heavy polypeptide chain with papain releases a fragment with molecular weight 37 000, presumably leaving an 8000 molecular-weight tail stuck in the lipid bilayer (figure 7.15).

The human HLA-A and HLA-B genes appear to be closely related to the H-2K and H-2D genes of the mouse, and the corresponding human antigens are under intense study. The general structure of figure 7.15 applies also to HLA-A and HLA-B antigens which consist of a trans-membrane glycoprotein of molecular weight about 44 000, non-covalently associated with β_2-microglobulin.

As is the case with most cell surface antigens, detailed structural studies on H-2 and HLA antigens are hindered by the necessity to purify specific polypeptide components from complex mixtures using only small

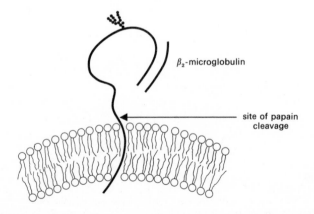

Figure 7.15. Schematic structure of H-2D, H-2K (heavy chain approx. 45 000) and HLA-A, HLA-B, HLA-C (heavy chain approx. 43 000) antigens.

amounts of membrane material. Progress in this area has been aided by the application of new techniques, possibly the most powerful of which is the production and use of *monoclonal antibodies*. Injection of mice with antigen stimulates the production of spleen cells, any one of which is primed to produce antibodies directed against a specific determinant on the injected antigen. Fusion of such spleen cells with myeloma cells (immunoglobulin-producing tumour cells) leads to a range of hybrid cells, each of which can produce relatively large amounts of a specific antibody. Selection and culture of an appropriate single hybrid cell can accordingly lead to a clone of cells producing monospecific antibodies on a reasonable scale. Such antibodies can then be used as affinity ligands (page 90) in rapid and relatively large-scale purification of the antigen molecules. Amino-acid *microsequencing* techniques have also been applied in the study of histocompatibility antigens. Mouse cells are grown in culture in the presence of a radioactively-labelled amino acid, which is consequently incorporated into all newly-synthesized proteins, including the required antigen. This is then isolated and sequentially degraded by standard techniques, except that the specifically-labelled amino acid can be detected with a sensitivity some million times higher than that of conventional sequence analysis. The entire procedure must be repeated many times with different labelled amino acids, and is accordingly laborious. Nevertheless, considerable progress has been made in the sequencing of H-2 antigens using these means.

In the case of human antigens, the use of large-scale cultures of cell lines carrying a high density of HLA antigens, together with affinity purification using monoclonal antibodies, has allowed the complete sequencing of certain HLA antigens by conventional methods. The primary amino-acid structure of HLA-B_7 heavy chain determined in this way shows many similarities to that of glycophorin A (page 95). Thus a hydrophobic sequence of 26 amino acids corresponds to that section of the HLA-B_7 antigens that crosses the membrane bilayer, and a C-terminal tail, 31 residues long and rich in hydroxyamino acids, extends into the cytoplasm. The extracellular N-terminal chain (271 residues) is much longer than that of glycophorin. Present data indicate considerable homology between the sequences of HLA-A and HLA-B on the one hand, and also between HLA and H-2 antigens on the other, while differences between the species seem to be concentrated in two defined extracellular variable regions. The HLA antigens appear to carry a single carbohydrate side-chain, N-glycosidically linked to an asparagine residue (see page 137) in the N-terminal peptide.

The association of heavy and light (β_2-microglobulin) polypeptide chains that make up the histocompatibility antigens is reminiscent of the serum immunoglobulins, the structures of which are well characterized. Moreover, present indications are that primary and secondary polypeptide structures of the two molecular classes also show many similarities. Immunoglobulins are found not only in serum but also membrane-bound to cell (lymphocyte) surfaces and, although the membrane-bound species have been less studied, the suggestion is of an evolutionary relationship between histocompatibility antigens and antibody (immunoglobulin) molecules.

The antigenic specificity of histocompatibility antigens seems to be carried by the heavy glycoprotein chain rather than by β_2-microglobulin, which could possibly serve to hold its larger neighbour in an antigenically-active conformation. It is currently believed that the antigenic specificity of the glycoprotein resides in its amino acid sequence, rather than in the carbohydrate side-chain which, at least in H-2 antigens, can apparently be destroyed without loss of antigenicity.

The mouse H-2 complex codes also for the Ia (I-region associated) antigens which are determined by the I-region (figure 7.14). Although five I subregions have been defined, two of these, I-A and I-C, code for most of the known Ia specificities; specificities from each I-subregion are expressed as separate molecules on the surface of cells, predominantly B lymphocytes. Methods similar to those employed in the study of H-2 antigens have indicated that Ia antigenic specificity on B lymphocytes is carried by two associated transmembrane integral glycoproteins, α (35 000) and β (25 000) (figure 7.16), both of which are believed to carry an N-glycosidically-linked oligosaccharide side-chain. The human equivalents of the mouse Ia antigens were studied at the 1977 (seventh) International Histocompatibility Workshop and were shown to be gene products of a locus closely related to that previously defined as HLA-D (figure 7.14). The locus for human Ia-type antigens has been provisionally termed HLA-DRw (meaning D-related workshop) and the corresponding antigens as HLA-DRw antigens. Like mouse Ia antigens, HLA-DRw antigens are composed of two apparently dissimilar transmembrane glycoproteins; the molecular weights in the human case being estimated at 33 000 and 28 000 respectively (figure 7.16). The individual-specific determinants of HLA-DRw antigens are believed to be located on the 33 000 molecular-weight polypeptide. Whether these determinants are carried by amino-acid sequences or by carbohydrate is, as in the case of mouse Ia antigens, unclear. The general opinion, based upon the

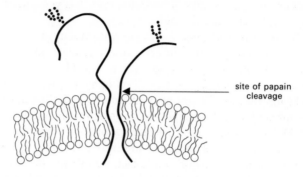

Figure 7.16. Schematic structure of mouse Ia (chains approx. 35 000 and 25 000) and human HLA-DRw (chains approx. 33 000 and 28 000) antigens: based on C. J. Barnstaple *et al.*, *Br. Med. Bull.* (1978), **34**, 241–246.

susceptibility of antigenic activity to destruction by heat and protease digestion, as well as upon its resistance to glycosidase treatment, favours the polypeptide structure as the antigenic determinant. However, individual sugars have been claimed to block anti-mouse Ia sera specifically, so implicating the carbohydrate components, and the question must remain open at present.

Differentiation antigens

Differentiating cells, particularly lymphoid cells, are known to carry surface *differentiation antigens* which change with, and can be used to follow, the differentiation process. For example, bone marrow precursors of mouse T lymphocytes acquire in the thymus the differentiation antigens Thy-1, TL, Ly-1 and Ly-2, 3, and subsequently lose TL antigens on further differentiation into mature T cells. Ly-1 and Ly-2,3 antigens appear to be markers for functional subclasses of T cells. Ia antigens, present on B lymphocytes but not, in general, on T cells, can also be classified as differentiation antigens but are usually, as in the present chapter, discussed in the context of the H-2 gene complex.

Thy-1 antigen (formerly called θ antigen) is expressed in large amounts on brain cells and thymocytes of mouse and rat, and is commonly used as a T-cell marker in mice. The antigen is an integral membrane glycoprotein which, in the case of rat brain and thymus, has been shown to have a molecular weight of approximately 18 500 and to contain 30% by weight of carbohydrate, N-glycosidically-linked to three separate asparagine residues.

TL (thymus leukaemia) antigens are present in high concentration on mouse thymocytes and on certain leukaemias, but not on B cells or on mature T cells. They are coded for by the TL locus, closely linked to the H-2 gene complex (figure 7.14). Preliminary biochemical characterization indicates that the TL antigens are similar to H-2D and H-2K antigens in comprising a heavy integral glycoprotein component (molecular weight 50 000) non-covalently associated with β_2-microglobulin.

Ly-1 and *Ly-2,3 antigens* are restricted to thymocytes and peripheral T cells, the majority of which express all three antigens Ly-1, Ly-2 and Ly-3. It seems that further functional commitment of these cells leads to restriction of the Ly antigens in that 'helper' T cells carry only Ly-1 antigens, whereas 'killer' or suppressor cells carry both Ly-2 and Ly-3 antigens. These latter antigens have so far only been found to be expressed together, and it has been suggested that they are coded for by a complex locus *Ly-2,3* and expressed on the same molecule. What little biochemical information is available in this system has been derived from the Ly-2,3 antigen which appears to be a glycoprotein with molecular weight approximately 35 000.

Since the development of affinity chromatography using monoclonal antibodies (page 151), progress in the isolation and characterization of lymphocyte differentiation antigens has been very rapid. A number of rat lymphocyte antigens [e.g. L-C (leucocyte common), W3/13, W3/15, W3/25 and OX-8] have been identified by these means, and it can be expected that many more will be defined in the coming years.

Whereas the carbohydrate content of some of these antigens is known to be high [e.g. Thy-1 (30%) and L-C (24%)], the precise roles played by their oligosaccharide residues in antigenic interactions are unclear and clarification must, as in the cases of H-2 and HLA-coded antigens, await more complete structural analysis.

Receptors for bacterial toxins and glycoprotein hormones

In addition to their role as cell surface antigens, membrane-bound glycolipids and glycoproteins have the capacity to act as receptors for a range of extracellular agents.

Cholera toxin is a protein of molecular weight 84 000 that binds to cell membranes of mammalian intestinal epithelium, so initiating changes in transport of salt and water that result in massive diarrhoea. The toxin comprises a B protein composed of five identical subunits and an A protein made up of two non-identical subunits, A_1 and A_2. The mechanism of action is believed to involve initial binding of the B chains to a cell membrane ganglioside G_{M1} (Table 5.4, p. 80) which serves as a specific receptor for the toxin. Binding inhibition studies show the toxin to have much higher affinity for G_{M1} than for other gangliosides shown in Table 5.4. Following binding, the B chains undergo a conformational change, releasing and 'activating' the A_1 subunit which translocates within the membrane and stimulates adenylate cyclase with production of cyclic AMP. The increased production of cyclic AMP is thought to promote changes in the permeability of the tight junctions between the cells, and results in a loss of salt and water into the intestinal lumen (see page 193).

Tetanus toxin is a neurotoxin that is presumed to block synaptic inhibition in the central nervous system, so leading to muscle rigidity, convulsions and occasionally death by respiratory failure. Much less is known about the toxin and its mode of action than about cholera toxin but, like the latter, it has been shown to bind specifically to a ganglioside component of its target membrane (in this case on nerve cells). In contrast to cholera toxin, tetanus toxin shows greatest affinity for the ganglioside G_{D1b} (Table 5.4). Still less is known about *Diphtheria toxin*, which has been shown, by hapten inhibition studies, to bind to carbohydrate-containing receptors on membranes of its target cells, which are apparently tissue non-specific.

Membrane-bound gangliosides are also involved in receptor sites for a number of glycoprotein hormones. Thus, binding of *thyroid stimulating hormone* (TSH), *human chorionic gonadotropin* (hCG) and *luteinizing hormone* (LH) to their target membranes is most potently inhibited by G_{D1b}, G_{T1b} and G_{T1b} respectively (Table 5.4) TSH is the best studied of these hormones and, like cholera toxin, contains two types of subunit, one of which (β) is believed to bind to a membrane receptor, following which the second subunit (α) translocates within the membrane and activates adenylate cyclase. Some remarkable similarities in amino-acid sequences occur between corresponding subunit types of TSH and cholera toxin (B and β, A and α) which also extend to the other hormones hCG and LH, suggesting some common mechanisms (see e.g. page 232) in the course of producing widely different end results. There are even some parallels with the antiviral agent *interferon* which binds to cell membrane-bound carbohydrates and causes raised levels of cyclic AMP in the course of its antiviral action.

Although there is strong evidence for the participation of specific gangliosides as membrane receptors for most of the above toxins and hormones, other molecules may be involved. Glycoprotein receptors for TSH and hCG have also been isolated, and it has been suggested that glycolipid and glycoprotein molecules might participate together in receptor function. It is, moreover, by no means certain that the gangliosides currently cited as binding most potently to the various toxins and hormones are necessarily the true physiological receptors. It is conceivable that minor uncharacterized membrane components might show even greater affinities, and some evidence for this has recently been reported in the case of TSH.

The membrane-bound receptors for other hormones and for some neurotransmitters [e.g. acetylcholine (page 217)] are known to involve glycoproteins and in many such cases the molecular basis of the binding specificity is not known.

Liver receptors for desialylated plasma glycoproteins, erythrocytes and lymphocytes

A further receptor activity of membrane-bound glycoprotein was recognized when desialylated ceruloplasmin (a normal plasma sialoglycoprotein) was intravenously injected into rabbits and found to be cleared from the circulation in a matter of minutes. Under similar conditions the intact sialoglycoprotein survived for days. Subsequent investigations showed the phenomenon to be general, in that a number of desialylated plasma proteins are similarly removed from the circulation of a range of mammalian species. The asialoglycoproteins are recognized and bound by the liver, where they are catabolized in the lysosomes. An integral glycoprotein receptor has been isolated from liver plasma membranes using affinity chromatography (page 90), and has been shown to be specific for terminal galactose residues (exposed by removal of sialic acid) on the bound desialylated plasma proteins. The purified receptor is a glycoprotein (10 % by weight carbohydrate) which aggregates in water. Digestion with the proteolytic enzyme mixture, pronase, allowed isolation of two distinct glycopeptide fractions for which the partial structures shown in figure 7.17 have been proposed. It is of interest that one of these structures has a 'core' pattern which, like those proposed for glycophorin-derived glycopeptides (page 148) differs from that of the general model shown in figure 7.7. It could be that the latter model, largely derived on the basis of secreted glycoproteins, is not typical of membrane-bound glycoproteins, of which few have as yet been fully characterized.

The asialoglycoprotein-binding protein can be classed as a mammalian lectin (page 131) as it is specific for carbohydrate (galactose) and, in its aggregated water-soluble form, can agglutinate red cells. The ability of the lectin to bind asialoglycoproteins seems to be destroyed by neuraminidase treatment and restored either by resialylation or by further degradation by galactosidase. In fact, it is likely that asialolectin recognizes and binds to its own exposed galactose residues, so inhibiting binding to other asialoglycoproteins. This inhibition is removed when the terminal galactose residues of the asialolectin are masked (resialylation) or cleaved (galactosidase).

A similar mechanism for removal of certain circulating glycoproteins appears to exist in chicken liver, from which an analogous glycoprotein receptor has recently been isolated and shown to bind specifically to

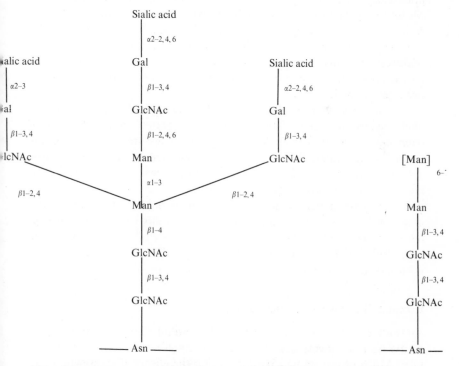

Figure 7.17. Partial structures proposed for the oligosaccharide complexes of two glycopeptides isolated from the asialoglycoprotein receptor of mammalian liver plasma membranes: from T. Kawasaki and G. Ashwell, *J. Biol. Chem.* (1976), **251**, 5292–5299.

glycoproteins carrying a terminal N-acetylglucosamine residue (structures from which both sialic acid and galactose have been removed).

Mammalian liver has the capability of recognizing specific determinants normally masked by sialic acid, not only on circulating soluble glycoproteins but also on intact cells. Thus, treatment of *erythrocytes* with neuraminidase and reinjection of the asialoerythrocytes into the circulation of the donor leads to their accelerated clearance and to their accumulation in the liver.

This phenomenon may well represent the normal physiological mechanism for the removal of senescent erythrocytes, which are known to contain relatively low amounts of sialic acid. Despite the obvious analogy with the removal of circulating asialoglycoproteins discussed above, the clearance of erythrocytes is in some ways different. Thus, whereas asialoglycoproteins are bound to liver hepatocytes, asialoerythrocytes apparently bind to Kupfer cells and also to mononuclear spleen cells. Moreover, galactose oxidase-mediated oxidation of terminal galactose residues of asialoglycoproteins prevents their removal from the circulation. This is not true of mammalian asialoerythrocytes which may well be removed by a distinct system. It should also be emphasized that the molecular nature of the hepatic receptor for asialoerythrocytes has not been determined.

A similar recognition of membrane-bound sugar determinants by an uncharacterized receptor has been demonstrated in the *homing of lymphocytes*. Radioactively-labelled lymphocytes transfused into rats normally accumulate in the spleen and lymph nodes within a few hours. Treatment of the lymphocytes with neuraminidase before transfusion, however, caused a rapid build-up of liver-localized radioactivity which decreased over a period of 48 hours, while that of the lymph nodes gradually increased to control levels. These events can be explained in terms of recognition and binding of desialylated lymphocytes by the liver, followed by gradual regeneration of cell surface sialic acid with release of the cells and migration to their normal target.

Receptor sites for viruses and bacteria

Viral attack on a mammalian cell involves initial attachment of the virus to the host cell membrane, followed by engulfment of the virus by the host. Much of our knowledge concerning the effects of viruses in mammalian systems is derived from studies of the *myxoviruses* [enveloped viruses (chapter 4) containing RNA] such as *influenza virus*. Influenza virus particles will bind simultaneously to a number of red blood cells causing haemagglutination, although, because the erythrocyte is not a permissive host, attachment in this case is not followed by engulfment. Nevertheless, the interaction has been extensively studied and is known to involve binding of viral envelope spikes (glycoprotein in nature) to specific receptor sites on the erythrocyte membrane. The erythrocyte receptors have been identified as glycophorin (page 95) molecules, and more specifically as sialic acid residues, as prior treatment of erythrocytes with neuraminidase (originally called 'receptor destroying enzyme') eliminates binding. Mammalian cell surface glycoproteins have also been implicated in the receptor sites for another enveloped virus *Semliki Forest virus* (SFV). In this case glycoprotein-containing spikes from the viral envelope have been shown to bind specifically to the major histo-compatibility antigens (page 149) of human and mouse cells in culture. The demonstrated homology of amino-acid sequences in human (HLA-A and HLA-B) and mouse (H-2K and H-2D) antigens (page 151) is consistent with the wide host range of SFV.

Bacteria also may recognize and bind to sugar-containing target sites on mammalian cell membranes. Several strains of *E. coli* have been shown to bind specifically to D-mannose residues on the surface of human epithelial cells. The binding agent is believed to reside in the pili that protrude from the bacterial surface, and a mannose-specific lectin has been isolated from *E. coli*. Similar evidence for the involvement of mannose in the attachment of *Salmonella typhi* to mammalian cell membranes has been obtained, and fucose has been implicated as a target site for *Vibrio cholera* on rabbit intestinal brush borders. The role of bacterial attachment in bacterial infection is not clear but, if attachment is a prerequisite to colonization, then it is possible that hapten inhibition of binding, using receptor-like substances, could be employed to prevent infection.

Cell-cell adhesion

Some examples of apparently specific interactions between different mammalian cells (e.g. erythrocyte-liver and lymphocyte-liver) have already been discussed (page 158). Adhesion between cells of the same type also occurs and is a fundamental prerequisite for morphogenesis and organogenesis. Present indications are that such adhesion, too, requires specific recognition processes, probably involving cell surface carbohydrate. Research in this area was greatly stimulated by a theory postulated in 1970 by Roseman, who proposed that cell-cell adhesion is mediated by enzyme-substrate recognition, in that a glycosyl transferase on the surface of one cell binds specifically to its substrate oligosaccharide on a neighbouring cell. Attractive as this proposal is, it has not so far received convincing experimental backing. A similar 'lock and key' mechanism involving specific linkage between lectins and sugar receptors is, on the other hand, receiving growing support. The isolation of a galactose-specific lectin from mammalian liver cells was discussed on page 156, and further examples have recently been reported, chiefly in non-mammalian systems. The cellular slime moulds, in particular the species *Dictyostelium discoideum*, have become popular model systems for eukaryotic development and have proved useful in cell adhesion studies. At a certain stage in their life-cycle, single slime mould cells adhere to each other forming a multicellular 'slug' or 'grex'. That the aggregation is specific is shown by the formation of two different homogeneous slugs from mixed aggregates of two closely-related species. Coincident with the development of cohesiveness, *D. discoideum* cells acquire cell surface proteins with lectin activity. The proteins have been purified to yield two lectins, discoidins I and II, with apparent specificities for *N*-acetyl-D-galactosamine and lactose respectively. The involvement of the lectins in the cell adhesion process is inferred from inhibition studies in which the endogenous cohesiveness of the slime mould cells is blocked by addition of the appropriate sugars. Similar lectin activities (pallidins) with apparently similar functions have also been isolated from another slime mould *Polysphondelium pallidum*.

Despite the progress made in the isolation of adhesive components from cellular slime moulds and also from sponges, parallel information concerning higher organisms remains sketchy. Galactose-binding lectins have been reported in a number of tissues including the electric organs of the electric eel and chick embryo pectoral muscle, but without strong evidence for their function. One possible example of a specific lectin-mediated adhesive mechanism comes from the chick retinotectal system. Ganglion cells from the retina send axons to the optic tectum, and inversion occurs such that cells from the dorsal half of the retina preferentially adhere to cells from the ventral half of the optic tectum and vice versa. It has been reported that the dorsal retina is rich in hexosamine-terminal carbohydrate, whereas the complementary ventral half of the tectum contains a protein that preferentially binds to *N*-acetyl-D-hexosamine.

Over the last few years a great deal of interest has been generated in the *fibronectins*, large extracellular proteins believed to be involved in cellular adhesion. Fibronectins from various sources probably represent just two specific proteins, cell surface and plasma fibronectins respectively. Cell surface fibronectin [also known as LETS (large external transformation-sensitive) protein (see page 162) or as CSP (cell surface protein)] is a glycoprotein of apparent subunit molecular weight 200 000–250 000 which can be detected (by fluorescent-labelled specific antibodies) in striking fibrillar arrays on the surfaces of many cultured cells. Fibronectin is apparently only loosely associated with the cell membrane, being readily detached by sucrose gradient centrifugation. It is not a conventional peripheral membrane protein in that it extends between cells and is also found sandwiched between cells and the tissue culture substratum. Such a localization is consistent with its presumed adhesive role, the assignment of which is largely based upon the ability of purified fibronectin to increase the adhesion of cells to each other and to the substratum. Plasma fibronectin [also known as Cig (cold insoluble globulin)] is a dimeric glycoprotein very similar to the cell surface form, and has a subunit molecular weight of 200 000–220 000. It is present in human plasma (0·3 mg/ml) and serum, and is believed to be the serum factor commonly required for the attachment of certain cells to collagen-coated dishes or for cells to attach and spread on tissue culture dishes. A further factor with molecular weight 140 000 has additionally been isolated from chick serum. Cell surface fibronectin is equally active in promoting cell spreading and attachment to collagen, and cells with a high level of the cell surface protein do not require serum factor.

The fibrillar organization of cell surface fibronectin can be shown by double-label immunofluorescence to correlate closely with the distribution of cytoplasmic actin microfilaments (page 119), and it has been suggested that the two fibrillar systems are interconnected, perhaps at plaques which mark points of cell-cell or cell-substratum contact. The intra- and extra-cellular systems could be linked via integral membrane proteins (cf. figure 6.16), but no hard evidence for specific linkage proteins has yet appeared. The nature of the immediate interaction between fibronectin and the cell surface is not clear, although fibronectin has been shown to be relatively ineffective in mediating cell-substratum adhesion of mutant baby hamster kidney (BHK) cells that are deficient in cell surface carbohydrate. This would seem to implicate carbohydrate-fibronectin interactions in the adhesive process.

Surface changes in cancer cells

Tumour cells have the capacity for uncontrolled growth and, in the case of malignant cells, migration out of tissue bounds, and it is generally assumed that these properties are related to changes in cell recognition and adhesive processes. As has been discussed earlier in this chapter, such processes appear to be controlled by cell membrane components, and the present section will outline briefly some of the best-characterized molecular differences between the cell membranes of normal cells on the one hand, and transformed or tumour cells on the other. It is worth clarifying the last two terms. *Transformed cells* are cells, mainly of fibroblastic origin, that have been changed *in vitro*, either spontaneously or by an oncogenic determinant (e.g. a virus or chemical carcinogen) so that one or more of certain properties are altered. Such changes include loss of contact inhibition of movement and growth, reduced serum requirement and loss of anchorage dependence. If transformed cells, when transplanted, give tumours in immunologically-impaired mice (e.g. nude mice) then they are classed as *tumour cells* as are, of course, cells of tumours that have arisen spontaneously *in vivo*. Not all transformed cells are necessarily tumour cells. A further distinction is made between *benign* and *malignant* tumour cells, based on a number of factors, including the more rapid growth of malignant cells and, more particularly, on their characteristic ability to invade and colonize both surrounding and distant tissue sites (*metastasis*).

Much of the current interest in fibronectin stems from observations that many transformed cells lack cell-surface fibronectin [hence its earlier name LETS protein (page 161)]. Moreover, addition of purified fibronectin to transformed cells has a number of 'detransforming' effects, such as increased adhesion to the substratum and assumption of more normal morphology, with less cell-surface microvilli and membrane ruffles. It is not clear exactly how a lack of fibronectin could contribute to the transformed phenotype. The extracellular matrix of normal cells contains not only fibronectin but also glycosaminoglycans (page 121) and collagen, all of which may be decreased in transformed cells. Fibronectin is known to bind to collagen (Latin: *fibra*, fibre; *nectere*, to bind) and it may be that a normal fibronectin-collagen-glycosaminoglycan matrix controls cell movement and growth. The transformed cell in culture is to some extent freed from external restraints of this type. Cell surface fibronectin is also apparently associated with development of a cytoskeletal network (page 161), and microfilament actin bundles are also decreased in transformation. A reduced fibronectin-cytoskeleton interaction could partially explain some properties of transformed cells, such as diminished cell adhesion and increased lateral mobility of integral membrane proteins (which could in turn explain the demonstrated increase in lectin agglutinability of transformed cells).

If, then, some of the observed characteristics of transformed cells could result from a decrease in cell surface fibronectin, what causes this decrease? There are a number of possibilities, but no clear answer. The synthesis of fibronectin could be decreased, or its proteolytic release from the cell surface could be increased; the decrease in cell surface fibronectin could result from an impaired microfilament function (although it is equally possible that microfilament malfunction could be caused by loss of cell surface fibronectin), or fibronectin receptors could be missing from the cell membrane. There is presently no convincing evidence for the generality of any of these factors, nor is fibronectin the only cell surface component for which changes have been reported in transformed and tumour cells. Both glycolipids and glycoproteins show transformation-sensitive changes, some of which will be briefly outlined here.

In 1968 a reported comparison of polyoma-transformed BHK fibroblasts with normal cells showed a remarkable decrease of the ganglioside G_{M3} (Table 5.4) and a corresponding increase in its biosynthetic precursor, lactosylceramide (figure 5.26), in the transformed cells. Since then, studies of many transformed and tumour cells both *in vitro* and *in vivo* have demonstrated a widespread simplification of membrane glycolipid compared with controls. Not only gangliosides but also neutral glycolipids and blood-group-active fucolipids (page 147) are effectively shortened by what is believed to be transformation-dependent inhibition of synthesis. Although common, such changes are not completely general, and a number of exceptions have been reported.

Changes in cell surface glycoprotein also have been associated with transformed and tumour cells. Gel filtration of glycopeptides proteolytically cleaved from the surfaces of a wide range of transformed cells has consistently shown the presence of a major fucose-containing peak that is eluted faster than the corresponding peak from control cells. Similar alterations in elution profiles have been demonstrated in tumour cells *in vivo*, and the general conclusion has been drawn that the changes are actually related to tumourgenicity rather than simply to the transformed phenotype. The fast-eluting glycopeptides are characterized by a relatively high content of sialic acid, and structural analysis of individual glycopeptides indicates that this may reflect increased branching (by sialic acid-Gal-GlcNAc chains) of the glycopeptide 'core' (figure 7.7).

Reports of apparently tumour-specific proteins are not uncommon, and just two further examples will be quoted here. Electrophoretic evidence has been cited for the presence in malignant tumour cells of membrane proteins that are apparently specific for the preferred target site of metastasis and whose amount increases with the cells' metastatic potential. Rather more dramatically, recent reports have described a single membrane glycoprotein with molecular weight 100 000 which is present in a characteristically-modified form (binding relatively high amounts of concanavalin A) in a wide range of tumour cells.

Chemical characterization of cell surface components represents one approach towards finding specific tumour cell markers which could not only serve a diagnostic purpose but might also act as targets for therapeutic measures. Cancer immunology represents another such approach, and indeed many tumour-specific antigens have been defined by immunological methods [e.g. TL antigens (page 154)]. A present hope is that, perhaps by a combination of the two techniques, tumour antigens might be characterized and used in the immunotherapy and possibly even vaccine-based prevention of cancer.

SUMMARY

1. The plasma membranes of animal cells contain glycoproteins and glycolipids which extend their carbohydrate portions into the extracellular environment. The presence of such cell-surface carbohydrate has been amply demonstrated by cytochemical (particularly using lectins) and electrokinetic techniques.

2. Glycolipids can be extracted from membranes by using polar organic solvents and are purified by standard techniques of lipid chemistry. Membrane glycoproteins are generally integral proteins and accordingly require solvents, detergents or chaotropic agents for solubilization. Fractionation of glycoproteins has been facilitated by the development of lectin affinity chromatography, although purification in conventional terms is greatly complicated by microheterogeneity.

3. The carbohydrate of membrane glycoproteins is contained in oligosaccharide side-chains attached to a polypeptide backbone by one of two major linkage types.

4. The oligosaccharide complexes of glycoproteins are biosynthesized by sequential transfer of monosaccharides from nucleotide precursors to a lipid carrier and are then transferred to the polypeptide chain where they are further modified. Glycolipid biosynthesis involves direct transfer of monosaccharides from sugar-nucleotides to the sphingoid of the final structure.

5. ABO and Lewis blood group activities are specified by terminal carbohydrate residues of plasma membrane glycolipid and glycoprotein, and their relationships have been

worked out by using soluble blood-group active glycoproteins. MN blood group activity is associated with the major sialoglycoprotein of the erythrocyte, glycophorin A and, while carbohydrate residues are essential for antigenic activity, the specificity is believed to lie in the amino-acid sequence.

6. Histocompatibility and differentiation antigens all appear to be integral membrane glycoproteins, a number of which have been purified and subjected to amino-acid sequencing. The structures and roles of the oligosaccharide side-chains have, in general, not been determined.

7. Cell surface receptors for a number of bacterial toxins and glycoprotein hormones have been characterized and shown to be glycolipids, whereas liver receptors for desialylated plasma glycoproteins are themselves sialoglycoproteins. The interaction between some viruses and their host cells also involves membrane-bound glycoprotein on the surfaces of both virus and host, while membrane-bound carbohydrates have been shown to act as epithelial cell receptors for a number of pathogenic bacteria.

8. There is growing evidence, particularly from the slime-mould and sponge systems, that cell-cell adhesion might be mediated by surface lectin-carbohydrate interactions. The extracellular glycoprotein, fibronectin, has also been implicated in cellular adhesion mechanisms.

9. Fibronectin is also of interest because of its apparent reduction on the surfaces of transformed cells, although the relationship of this reduction to the transformation process is unclear. Glycolipids of transformed cells tend to be less complex than those of normal controls, whereas glycoproteins tend to be more complex in the former. Despite many reports of tumour-specific antigens, no single general species of this type has yet been fully defined.

CHAPTER EIGHT

MEMBRANE TRANSPORT MECHANISMS

AS WE SAW IN CHAPTER TWO, A FUNDAMENTAL FEATURE OF MEMBRANES IS a variably-restricted permeability to many of the solutes found in the surrounding aqueous environment. The movement of molecules across membranes may be the result of simple diffusion, or it may involve a complex integration of membrane functions with the energy-yielding metabolic processes of the cell. The molecular nature of both simple and complex transport processes is what concerns us in this chapter. Much of our current knowledge stems from studies on prokaryotes but, as in many other areas of biochemical research, in the absence of evidence to the contrary the molecular mechanisms being elucidated are put forward as models for all biological systems.

Diffusion

Within the bulk phase of a solution, random movement of solute molecules ensures their distribution throughout the solution. Should concentration differences be introduced, such random movements will lead to a net movement of solute from higher concentration areas to lower concentration areas, so that a uniform distribution is once more attained. The movement of molecules under these circumstances is termed *diffusion*, and tends to a state of maximum entropy (i.e. minimum order). The rate of diffusion is directly proportional to the concentration gradient and is described by Fick's law:

$$J = -D\frac{dc}{dx}$$

where J = rate of movement or flux, D is the diffusion coefficient and dc/dx is the concentration gradient.

The introduction of a barrier, such as a membrane, into the solution may hinder the free movement of solute molecules. When considering the movement of solute molecules between membrane-separated compartments, it is therefore necessary to take account not only of the concentration gradient of the solute but also of the *permeability* of the membrane to the solute. We saw in chapter 1 that for some molecules the permeability is directly related to the oil-water partition coefficient of the molecules. The implication of this observation is that such molecules cross membranes by diffusing through the hydrophobic region of the fatty acyl chains which constitute the bulk of the membrane bilayer. Other molecules, however, do not behave in this simple manner. Water and many small compounds, e.g. methanol, urea, formamide and ethylene glycol, cross membranes much more rapidly than would be expected from a consideration of their oil-water partition coefficients. There is a considerable body of evidence that biological membranes are traversed by water-filled pores or channels, and experiments with black lipid films indicate that the presence of proteins in the lipid bilayer enhances the permeability of the membrane to water and to a variety of ions. Thus intrinsic membrane proteins may permit the formation of water-filled channels across the membrane.

Measurement of the osmotic pressure generated across membranes can give information about the size of the water-filled pores. The osmotic pressure of a solution of known solute concentration can be calculated on the assumption that the membrane system is permeable to solvent but not to solute. In practice, the observed osmotic pressure is frequently seen to be less than the calculated value, indicating that the membrane has permitted the passage of some solute molecules. The ratio between the observed or apparent osmotic pressure and the calculated value is termed the *reflection coefficient* and is a measure of the permeability of the membrane to solute. If the membrane is completely impermeable to solute, the apparent osmotic pressure is equal to the calculated value, and hence the reflection coefficient is 1, whereas a value less than 1 indicates a degree of permeability of the membrane to the solute. Comparison of the reflection coefficients of several compounds of known molecular radius therefore can provide a measure of the radius of the water-filled pores, and figures between 0·4 and 0·55 nm are commonly found. It is emphasized that these pores are not well-defined structures; they constitute a working concept which accommodates the observed behaviour of water and some small solute molecules.

Facilitated diffusion

The transport of most solutes across membranes involves processes other than simple diffusion. Thus many compounds show a much higher rate of passage than would be predicted from a consideration of their concentration gradient, their oil-water partition coefficient and $\Delta\psi$ (page 11). Furthermore, the activation energy is frequently considerably lower than would be expected for passage of the molecules through the membrane by simple diffusion. The major difference, however, is that Fick's law no longer holds, and the flux does not continue to increase with increasing solute concentration. Instead, the system shows *saturation kinetics*, i.e. increasing the concentration gradient increases the flux up to an asymptotic value. A graphical plot of transfer rate against concentration has the hyperbolic form of the familiar Michaelis-Menten curve. The implication of these observations is that transport of solute molecules is facilitated by their interaction with a finite number of specific membrane components. Such a conclusion is supported by further experimental evidence on the specificity of transport, which shows that optical isomers or close structural analogues of the transported solute may show greatly reduced rates of transfer. By analogy with enzyme systems, the interaction between the solute and the membrane component is described by an affinity constant K_m, and the rate of transfer or flux is analogous to V, the velocity of the enzymic reaction. Therefore the flux of solute in a facilitated diffusion system can be described by an equation of the same form as that of the Michaelis-Menten equation:

$$J = \frac{J_{max} \cdot [S]}{K_m + [S]}$$

where J = flux (as in the Fick's law equation),
$\quad J_{max}$ = maximum flux, i.e. the asymptotic value,
$\quad [S]$ = solute concentration,
$\quad K_m$ = affinity constant.

The same graphical treatments that are applied to data from enzyme kinetic experiments are applicable to transport data. Thus compounds thought to compete for the transport system may be characterized from Lineweaver–Burk plots of flux measurements made in the presence and absence of the putative competitor. In this way we can obtain a very clear picture of the behaviour of the transport system, how specific it is, the maximum transport rates that are achieved, and its interactions with a range of competitive and non-competitive inhibitors. However, such kinetic experiments tell us nothing of the molecular nature of the transport systems. The available evidence points to membrane proteins being the key elements. Once again, drawing our analogy with enzymes, only proteins offer the structural refinements necessary to provide highly specific interaction sites able to distinguish between, for example in the human erythrocyte, D- and L-glucose, the D isomer having a K_m over 1000 times lower than that of the L form, or between D-glucose and D-galactose, where a 10-fold difference in K_m is seen, D-glucose being the lower. Further confirmation of the role of proteins comes from experiments in which the effects of known protein-modifying reagents, such as mercaptoethanol, dithiothreitol, mercury ions, dinitrofluorobenzene and dansyl chloride are all seen profoundly to affect facilitated diffusion systems. There is, indeed, little doubt that the major constituent of each of the several facilitated diffusion systems found in most membranes is an integral protein.

If membranes contain specific transport proteins, would it not be possible to isolate such a protein and to study in detail the interaction with its substrate? Many research groups are trying to do this, but so far no major success has been reported. The literature contains abundant references to binding proteins isolated from many different cell types. These proteins usually show specific high-affinity binding for molecules that can be shown by kinetic studies to cross the membrane by a specific facilitated mechanism. However, with one or two exceptions, the experiments have not provided unequivocal evidence that the binding protein is primarily responsible for the transport of the molecule across the membrane. Experiments with proteins from erythrocytes do, however, go a long way towards this. Integral membrane proteins can be isolated by dissolution of the erythrocyte membrane with detergent, and one such protein (band 3 polypeptide, page 98) shows binding of anions. More importantly, it is able to confer specific anion transport properties to phospholipid vesicles and black lipid films (see page 104). Enzymic cleavage of the protein produces fragments which retain the anion-binding properties but have lost the transport capability. A similar approach has allowed identification of another protein (band 4·5, but see page 99) believed to be responsible for the facilitated diffusion of monosaccharides across the membrane. Although still somewhat controversial, this work illustrates the goal at which all transport research aims, i.e. the complete identification of the transporter, and its fully functional reconstitution into a chemically-defined lipid membrane.

We shall accept that the transporter systems are proteins, even though we cannot yet characterize them completely; the next question then is, how could these proteins be arranged in the membrane?

There are two mechanistic models which are fundamentally different and which represent the two extremes of current views. The first of these is the *mobile carrier hypothesis*. The transport protein is seen as a highly mobile membrane protein able to diffuse rapidly from face to face of the membrane. It has a specific high-affinity binding site for the substrate which can be exposed at either face of the membrane. Such a scheme is outlined diagrammatically in figure 8.1.

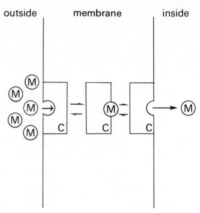

Figure 8.1. The mobile carrier hypothesis: metabolite M binds to mobile carrier C at the outer face of the membrane. The CM complex diffuses through the membrane and dissociates at the inner surface, releasing M into the cell interior.

The second interpretation, which may be termed the *pore hypothesis*, sees the transport system as a specific pore or channel through the membrane. Unlike the water-filled pores discussed on page 167, these pores are protein-lined and are permanent structures. Specificity may arise from pore diameter, charge distribution within the pore, the presence of specific binding sites at the pore entrance/exit or indeed combinations of all these factors (figure 8.2).

To discriminate between these models experimentally is not easy, and proponents of either model readily find fault with the interpretations of the opposing camp. There are certain basic characteristics of facilitated diffusion systems which either model has to accommodate. For example, it is seen experimentally that even when such a transport system has reached equilibrium, addition of a radiolabelled substrate to either side of the membrane results in a rapid appearance of labelled molecules on the other side, i.e. there is a unidirectional flux of labelled material through the membrane but, as there is an exactly equal and opposite flux of unlabelled material, the equilibrium is unchanged. This indicates that at equilibrium the transport system continues to operate, resulting in a rapid exchange but no net flux. Either the mobile carrier or the current pore model shows behaviour of this type.

We can find natural compounds that provide us with working models of both mobile carriers and pores. In 1964 C. Moore and B. C. Pressman

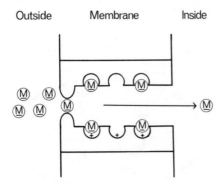

Figure 8.2. The pore hypothesis: trans-membrane proteins constitute a specific channel or pore across the membrane. Entry of metabolite M into the channel may depend on the size of M, and the progress of M through the channel may depend on charge distribution and/or the presence of specific binding sites in the channel.

discovered that an antibiotic *valinomycin* shows great specificity for the transport of potassium into mitochondria. Since that time, other naturally-occurring antibiotics have been found to function as highly-specific mobile carriers for alkali metal cations. Other compounds, also of bacterial origin, are able to function as ion-selective pores when introduced into phospholipid bilayers. These compounds are all classified under the generic name *ionophores*. They are a heterogeneous group with respect to molecular structure, but the cation carriers have the common property of being folded in such a way as to produce a hydrophobic exterior surrounding an interior cavity lined with critically-orientated oxygen atoms (see figure 8.3). These oxygens can bond with cations via ion-dipole interactions, in the same manner as a hydration shell forms around the ion in aqueous solution.

Some of the carriers form $1:1$ complexes with their respective alkali metal cation, and show marked discrimination between closely-related ions. Thus valinomycin shows a $10\,000:1$ preference for K^+ over Na^+; the difference in radius of the two ions is not great (K^+ being 0.133 nm

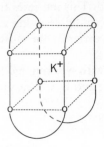

(*a*) Nonactin-K$^+$ complex;

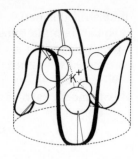

(*b*) Valinomycin-K$^+$ complex;

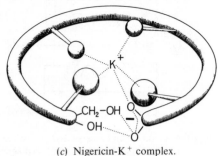

(*c*) Nigericin-K$^+$ complex.

Figure 8.3. Schematic representation of ionophore-cation crystalline complexes: (*a*) In the nonactin complex the potassium ion is held in quasi-cubic symmetry; (*b*) Valinomycin forms a complex in octahedral co-ordination with the potassium ion; (*c*) The potassium-nigericin complex shows no such symmetry. Hydrogen bonds hold the nigericin in a ring-like structure and the potassium is co-ordinated by the oxygen atoms projecting into the ring.

and Na$^+$ 0·095 nm) which serves to emphasize the precision of fit of the ion into the polar oxygen-lined cavity of the ionophore. There is unequivocal evidence that valinomycin behaves as a mobile carrier, ferrying K$^+$ ions across the membrane. There is a linear relationship between ion flux and ion concentration up to a point where saturation occurs. However, flux drops to zero if the temperature of the membrane is lowered to that of the phase-transition of the acyl chains (see chapter 6), indicating that a fluid membrane is a prerequisite for transport.

Other ionophores are equally clearly not mobile carriers. Gramicidins (A, B and C) and alamethicin are the two best-characterized *channel* or *pore-forming* ionophores. As in the case of the mobile carriers, these molecules present a structure with a hydrophobic exterior and an oxygen-lined central channel. The peptide chain of gramicidin A appears to dimerize in the membrane, forming a spiral about 3 nm long with a lumen 0·4 nm in diameter. Such a structure would span the hydrophobic region of a membrane and could provide a water-filled ion-conducting channel (figure 8.4). In general, the gramicidins allow the free passage of water and do not discriminate between ions to anything like the extent of the mobile carriers. Thus they can act as facilitated diffusion systems for a range of ions.

Alamethicin presents a more complex picture in that channel formation is seen to be dependent on $\Delta\psi$. Under conditions of zero $\Delta\psi$, alamethicin does not enhance ion permeability. However, as the voltage is increased, a point is reached when a water-filled ion-conducting channel is formed. The conductance is not related directly to alamethicin concentration, but to the fifth and ninth power of the concentration, suggesting that channels form by the aggregation of five or nine molecules of alamethicin within the membrane, and that this aggregation is dependent on $\Delta\psi$.

E. coli produces a group of proteins that have bactericidal activity and are collectively termed the *colicins*. The colicins have recently been shown also to form voltage-dependent water and ion channels in bilayer membranes. In the case of the colicins, a single protein molecule of molecular weight 50 000–100 000 is thought to form one channel without the need for aggregation, and it is possible that the voltage may serve to orientate the protein in the membrane. The existence of these systems has important implications for the electrically-excitable membranes that are found in nerve cells and muscles, and which we shall discuss in chapter 9.

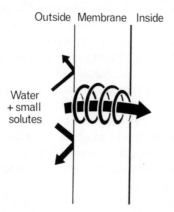

Outside Membrane Inside

Water
+ small
solutes

Figure 8.4. The gramicidin channel: the dimeric gramicidin forms a spiral peptide chain the lumen of which may provide a water-filled channel across the hydrophobic region of the membrane.

The polyene antibiotic, nystatin, typifies a further group of channel-forming antibiotics. In this case, an interaction with membrane cholesterol occurs, producing a water-filled membrane channel showing some preference for anions over cations. Measurements of the reflection coefficient for nystatin-containing membranes indicates that the diameter of the channel is of the order of 0·4 nm, a figure not very different from that of the diameter of the hypothetical pores discussed previously. Indeed, there is a suggestion that ionophores of the nystatin type induce the spontaneous formation of transient lipid micelles, leaving 'gaps' in the bilayer. The conductance increases markedly at low temperatures, which by increasing the rigidity of the acyl chains of the bilayer, would tend to stabilize the pores.

The ionophores provide models for both mobile carriers and channels in physiological transport systems. However, as stated at the outset, opinion is still very much divided as to which mechanism prevails. The last few years have seen a shift of opinion which tends to favour a channel as the most satisfactory explanation for facilitated diffusion. Examination of the pertinent literature, with its discussions of gated channels and moveable binding sites, may lead to the conclusion that the controversy is of a semantic nature rather than about fundamentally different molecular mechanisms.

Active transport

The foregoing discussion has been concerned with the mechanisms by which molecules may cross membranes under the influence of a concentration gradient. However, many transport processes work against a concentration gradient; in addition to the transporting systems discussed previously, such mechanisms require an input of energy. In several respects active transport processes resemble facilitated diffusion systems; thus saturation kinetics are seen, and there is evidence that specific interaction between substrate and membrane proteins is a critical step in the transport process. Although much of our knowledge of the detailed mechanisms of active transport processes derives from experiments with microbial cells, we can also find many well-characterized active transport systems in higher organisms.

Uptake of nutrients by the intestinal epithelial cells, absorption of nutrients by the liver cells from the blood, and re-absorption of essential metabolites from glomerular filtrate by the kidney tubule cells, all involve the transport of materials against the prevailing concentration gradient.

The pumping of ions, particularly K^+ and Na^+, into and out of cells against considerable concentration gradients is a feature of all plasma membranes, whether prokaryotic or eukaryotic. There is no single mechanism for the coupling of energy to transport systems. There are, however, fundamental concepts which appear to be relevant to all membrane-mediated active transport systems.

Group translocation systems

Through the late 1950s and early 1960s, Peter Mitchell expressed the view that membrane-bound enzymes could assume a vectorial character (see chapter 2) such that the chemical modification of substrate could proceed concomitantly with transfer across the membrane, and product could be released into a different compartment from that containing substrate. Such a process was described by Mitchell as *group translocation*, i.e. a chemical group (part of the substrate) is transported by the enzyme. It is important to distinguish between a true group translocation process and a system bearing a superficial resemblance to it. Thus, if a solute enters a cell by a simple diffusion process and is modified by an intracellular enzyme, then an apparent active accumulation of substrate is observed. The substrate is in effect removed or 'trapped', thereby maintaining a trans-membrane concentration gradient, and diffusion of substrate down its concentration gradient continues. In this case, unaltered substrate enters the cell, and it is in this feature that such a system differs fundamentally from a group translocation process, where chemical modification of the subtrate is an obligatory step in the transport process. Mitchell described a series of steps by which group translocation proceeds:

1. Interactions between solute and the water of the external environment are replaced by interactions between the solute and specific membrane constituents. It is assumed that no major change in free energy occurs at this step.
2. Bonds are broken within the solute molecule, releasing the group which is to be transported through the membrane.
3. New bonds are formed between the group and an acceptor.
   Steps (2) and (3) may be concomitant with the movement of the group through the membrane.
4. The modified group is re-hydrated and released at the cytoplasmic face of the membrane.

There are several bacterial transport systems which have most of the characteristics of a group translocation process and one of these, the *phosphoenolpyruvate phosphotransferase* system for sugar uptake, we shall now discuss in detail.

Many bacteria transport sugars by a process which leads to the accumulation of sugar phosphates in the cell. In 1964 Kundig and his colleagues discovered that an enzyme system associated with the plasma membrane of *E. coli* could transfer the phosphate group of phosphoenol-pyruvate to the C-6 hydroxyl group of hexoses. Since then the system has been studied extensively, and the details of the mechanisms are now clear.

In essence, the chemical energy available in phosphoenolpyruvate (PEP) is utilized to phosphorylate a transported sugar, but this is not a simple one-step transfer of the group directly to the sugar. In most organisms, four proteins are involved in the transport process. Two of these, Enzyme I and 'histidine protein' (HPr) are general proteins common to all phosphotransferase systems. The other two proteins are specific for particular sugars and are inducible; one of them (Enzyme II) appears to be a membrane-bound protein, whereas the second protein, Enzyme III, may be either membrane-bound or soluble. The sequence of reactions in which the four proteins participate is as follows:

Enzyme I has been partially purified from several organisms and has a molecular weight of about 70 000. The isolated enzyme preparation shows auto-phosphorylation using ^{32}P-PEP and the reaction is seen to require Mg^{2+}. HPr is a much smaller protein (~ 9000) which is readily purified and has been studied in a wide range of organisms. In every case, a single phosphate group is transferred to a histidinyl residue of the protein, when this is incubated *in vitro* with ^{32}P-PEP and Enzyme I.

The sugar-specific proteins, enzymes II and III, are not quite so well characterized. It seems that the two enzymes function together and that enzyme III should perhaps be considered as a cofactor or sub-unit of the enzyme II complex. The complex has a requirement for a specific phospholipid, phosphatidylglycerol, a fact which emphasizes the influence that membrane lipids may have on the activity of integral proteins (cf. page 117).

Discussions of the mechanisms and role of the phosphotransferase system inevitably raise the fundamental question—is the phosphorylation of the sugar an inseparable part of the transport process? Putting the · question another way—do enzymes II and III behave as facilitated diffusion carriers for free sugars which are then phosphorylated inside the cell by the cytoplasmic enzyme I—HPr system? There have not yet emerged unequivocal answers to these questions. Roseman's group has carried out an extensive study of the role of the phosphotransferase system in sugar transport. They find that *Staphylococcus aureus* mutants lacking enzyme I cannot accumulate either lactose or glucose analogues, both of which are normally accumulated as phosphate esters. These experiments were done with radiolabelled sugars, and suggest that no facilitated diffusion of the sugars into the cells occurs. Mutants lacking enzyme III of the lactose system were able to transport glucose normally, even though lactose uptake did not proceed. Kaback has carried out rather different experiments which complement Roseman's work. Kaback has shown that plasma membrane vesicles prepared from *E. coli* can accumulate glucose as its phosphate ester, whereas vesicles prepared from enzyme I-deficient mutants fail to do this. Using an elegant double-radiolabelling procedure, Kaback has also shown that external free sugar is more rapidly phosphorylated than internal free sugar, suggesting close coupling between sugar transport and phosphorylation.

There are, however, reports of experiments that indicate a possible facilitated diffusion role for the enzyme II complex of the transferase system. The experiments have been carried out on mutant strains of several different organisms. In a strain of *Salmonella typhimurium* it appears that galactose entry is mediated by an enzyme II complex, but that this carries the free sugar and is not coupled to a phosphorylation step. Similar studies have been made by Sir Hans Kornberg and Claudia Riordan using an *E. coli* strain which also shows uptake of free galactose by what is presumed to be an enzyme II complex; again no phosphorylation of the sugar occurs. Kornberg and Riordan have proposed a model in which the binding, translocation and dissociation of sugar are mediated by membrane constituents, but in which the phosphorylation is the responsibility of a cytoplasmic sugar kinase activity. Thus a facilitated diffusion system coupled to a trapping reaction appears overall as a vectorial-phosphorylation group-translocation process. Such a system is known to mediate the uptake of glycerol into several bacterial cells in which it accumulates as *sn*-glycerol-3-phosphate. Close examination reveals that the glycerol enters the cell by a facilitated diffusion process, but once inside it is immediately phosphorylated by an ATP-dependent glycerol kinase. In this case, accumulation of glycerol beyond the equilibrium point for diffusion is achieved by trapping, and there is no suggestion that the phosphorylation is vectorial.

With respect to the PEP system, however, the majority opinion at present favours a mechanism in which phosphorylation is tightly coupled to sugar translocation. Other examples of group translocation transport systems are also found in bacteria. Klein has shown that fatty acid uptake in *E. coli* is mediated by an inducible acylCoA synthetase that becomes membrane-bound in the presence of CoA. The transport of purine and pyrimidine bases into several species of bacteria has also been shown to be associated with a membrane-bound phosphoribosyl transferase.

We cannot yet formulate detailed mechanisms for the vectorial metabolism of sugars by the phosphotransferase system. It is clear, however, that a comprehensive picture of the individual constituents of the process is building up and that an understanding of the system at the molecular level is fast approaching.

Whereas the PEP system effectively couples the intrinsic energy of PEP to the transport of sugars, other transport processes utilize the energy gained from hydrolysis of ATP. The mechanisms used are quite different, however, and do not involve group translocation. The best characterized of the ATP-dependent systems are those concerned with ion pumping.

The Na^+, K^+-ATPase

In chapter 2 we saw that most membranes have an electrical gradient or *membrane potential* $\Delta\psi$ which arises from an unequal distribution of charged species across the membrane. As some of these species are large non-permeant molecules such as proteins and nucleic acids, mobile ions which may be able to permeate the membrane to a variable degree will tend to distribute themselves in such a way as to achieve overall electrical neutrality. In a hypothetical case, with a membrane completely permeable to ions, a *Donnan equilibrium* would be established. This is an equilibrium state in which an unequal distribution of ions is present but electroneutrality is achieved. In biological systems this simple situation does not exist; cell membranes are generally impermeable to ions, and measurements of $\Delta\psi$ for the plasma membrane in a variety of cell types give values ranging from -15 mV to -250 mV, the inside of the cell being negative with respect to the external medium. Thus, in considering the movement of ions across plasma membranes, we must consider not only the concentration gradient of the ion but also $\Delta\psi$.

In order to assess the relative contribution of these two parameters, it is convenient to express them both in the same terms. This is achieved by using the *Nernst equation* which relates the concentration of an ion to the voltage resulting from its concentration gradient. It has the following form:

$$E = \frac{RT}{FZ} \log \frac{[I]_o}{[I]_i}$$

where E = the electrical gradient in millivolts,
  R = the universal gas constant,
  T = the temperature in K,
  F = the Faraday (96·5 coulombs/mol),
  Z = the valence of the ion,
  $[I]_o$ = the concentration of the ion outside the cell,
  $[I]_i$ = the concentration of the ion inside the cell.

The equation can be simplified by converting the natural logarithms to base 10, when, for a univalent ion at a temperature of 20 °C, we have

$$E = 58 \log_{10} \frac{[I]_o}{[I]_i}.$$

The substitution of measured values of $[I]_o$ and $[I]_i$ into the equation gives a value in millivolts for E, the *equilibrium potential* for the ion, which represents the electrical gradient exactly corresponding to the chemical concentration gradient for that ion. Taking as an example the nerve cell, it is found that E for potassium ions is of the order of -65 millivolts to -70 millivolts compared to a measured $\Delta\psi$ of about -55 millivolts. Therefore the potassium ions inside the cell are subject to a net force of 10 to 15 millivolts which tends to drive them out of the cell. If we make the same calculation for sodium, a value of about $+50$ millivolts is found for E, thus an extracellular sodium ion experiences a net force of around 100 millivolts tending to drive it into the cell.

It can be demonstrated that both potassium ions and sodium ions diffuse in the directions expected from a consideration of E and $\Delta\psi$. However, it is also equally demonstrable that, in spite of such diffusion, the intracellular concentrations of these two ions, and indeed of all other ions, remain constant. Therefore there must be mechanisms for moving ions against the combined $\Delta\psi$ and concentration gradient of the ion. Such mechanisms, in thermodynamic terms, involve a decrease in the entropy of the system and therefore require an input of energy. In the early 1950s, R. D. Keynes and his colleagues showed that potassium ions are in constant flux across the plasma membrane. They demonstrated that movement out of the cell is independent of any metabolic energy supply, but movement into the cell is energy-dependent. Thus the passive outward movement of K^+ ions down their concentration gradient is counterbalanced by an active transport of ions into the cell. Keynes and his colleagues did these early experiments using the large axons of crabs and cuttlefish, but their findings are relevant to all plasma membranes. Hodgkin and Keynes considered the sodium ion and found that exit of Na^+ from the cell was energy-dependent, whereas Na^+ entry was not. They went on to show that the Na^+ and K^+ systems are linked, and that raising or lowering the external potassium concentration results in a parallel increase or decrease in the active outward flow of sodium ions.

In 1960 Caldwell and collaborators produced increased efflux of sodium from cyanide-poisoned squid axons by direct injection of ATP, GTP, arginine phosphate or phosphoenolpyruvate. Later work established that the active pumping of both sodium and potassium required ATP. Earlier, in 1957, Skou had discovered an ATPase in the leg nerve of a crab which showed a specific requirement for sodium and potassium. This finding was soon followed by reports of similar ATPases from a range of tissues, among which were cerebral cortex of guinea pigs, frog toe muscle, toad urinary bladder, seagull salt gland, and the electric organ of the electric eel. All these ATPases, usually referred to as Na^+, K^+-ATPases were found to be inhibited by a group of compounds called cardiac glycosides. These compounds had been shown to be specific inhibitors of the active transport of sodium and potassium by Schatzmann in 1953, and the idea developed that the Na^+, K^+-ATPase and the active ion pump were one and the same. It is now generally accepted that this is so. Much of the early evidence came from studies on erythrocyte ghosts and this is summarized below.

In intact ghosts, the pump and the Na^+, K^+-ATPase require the simultaneous presence of both K^+ and Na^+ and the effect of changing the concentration of either ion is the same on the pump and on the ATPase. It can be shown that Na^+ activates both the pump and the Na^+, K^+-ATPase only from the inside of the ghost, and the same is true of ATP. The cardiac glycosides inhibit both the pump and the Na^+, K^+-ATPase only from the outside, and the effects on both can be reversed by raising the external potassium concentration. We might expect that the Na^+, K^+-ATPase could run backwards thereby synthesizing ATP. This has been demonstrated and ^{32}P incorporation into ATP has been seen in the presence of abnormal cation gradients. The

ATP synthesis is also sensitive to cardiac glycosides. Consideration of the results of the many experiments briefly summarized above, leads to the conclusion that the Na^+, K^+-ATPase constitutes the Na^+, K^+ pump. The final goal of the researcher into transport systems is to reconstitute the physiological transport system in a completely defined model. This has been achieved for the Na^+, K^+-ATPase by Hokin's group. Hokin and his colleagues have been concerned with the purification to homogeneity of the Na^+, K^+-ATPase from two sources, the electric organ of the electric eel and the salt gland of the spiny dogfish shark. The group set out to incorporate their pure enzyme into a phospholipid bilayer. The enzyme was incorporated into single bilayer vesicles of phosphatidylcholine (see page 104) and the vesicles were extensively washed. Addition of ATP to the external medium stimulated $^{22}Na^+$ uptake into the vesicles against the concentration gradient. This was the 'opposite' to a normal cell where ATP is internal and Na^+ is pumped out. However, the group asked the question—if the vesicles pump Na^+ in, will they pump K^+ out? Vesicles were prepared with ^{42}K inside (or in most experiments ^{86}Rb, as ^{42}K is a very short-lived isotope) and an ATP-dependent exit of K^+ (or Rb^+) was seen. The transport of K^+ was seen to be dependent on external Na^+. The cardiac glycoside ouabain inhibited transport when present on the inside of the vesicles. Thus the reconstituted vesicles behave in almost all respects like a natural cell membrane, i.e. they will transport both Na^+ and K^+ against a concentration gradient in an ATP-dependent manner; the ATP must be on the same side of the membrane as the Na^+, and the system is inhibited by cardiac glycosides acting at the same side as potassium. It is not at all clear why the Na^+, K^+-ATPase inserts into the bilayer vesicles in the reverse manner from that found in natural membranes but, apart from this one aspect, the Na^+, K^+-ATPase appears to have been fully reconstituted as a trans-membrane pump. The stoichiometry of transfer is $2 \cdot 8$ Na^+ per 2 K^+, a figure remarkably close to that $(3:2)$ found in the erythrocyte.

There is no doubt that the Na^+, K^+-ATPase is the Na^+, K^+ pump. What then are the molecular mechanisms for the coupling of ATP hydrolysis to the trans-membrane movement of Na^+ and K^+? The ATPase has two types of sub-unit, a large polypeptide of molecular weight 85 000–100 000, and a smaller sub-unit with molecular weight about 50 000. This basic arrangement holds true for enzymes isolated from a variety of different sources. The larger peptide seems to span the membrane, a conclusion drawn from several quite different pieces of evidence.

Firstly, as first suggested in 1960 by Skou, ATP hydrolysis and cation transport are accompanied by a cyclic phosphorylation and dephosphorylation of the enzyme itself. The site of phosphorylation is an aspartyl residue on the larger peptide, indicating that a portion of the peptide chain must be exposed at the cytoplasmic face of the membrane. However, the peptide also has a binding site for ouabain which must be at the outer face of the membrane. Amino-acid analysis of the peptide reveals a high proportion of non-polar amino acids, suggesting that it may have a central hydrophobic region that can span the membrane (see page 92). The role of the smaller peptide is not clear. It is glycosylated, which would tend to suggest that it is located at the outer membrane surface (see page 93); this conclusion is supported by the fact that antibodies directed against it inhibit ATPase activity in intact erythrocytes. Rather surprisingly, antibodies against the larger chain have been reported to bind but not to inhibit ATPase activity.

The membrane-bound enzyme has a molecular weight of about 300 000, suggestive of a tetramer of two large and two small peptides. Early models of the pump assumed an asymmetric distribution of the sub-units within the membrane. Thus the *internal transfer model* of Lieb and Stein proposed that two of the larger sub-units were exposed at the cytoplasmic face of the membrane, and two of the smaller peptides carried the outward-facing K^+ and ouabain sites. The two cations, Na^+ and K^+, were transferred from one sub-unit type to another within the membrane—hence the name of the model. The Lieb and Stein model certainly accommodated most of the contemporary information about the pump. Thus it was known that for transport to proceed, binding of both Na^+ and K^+ must occur simultaneously—indicating simultaneous exposure of binding sites at both sides of the membrane. However, the binding constant for each ion is not affected by the concentration of the alternative ion, i.e. the binding constant for internal Na^+ is not affected by external K^+, strongly suggesting that the two ions bind to quite separate sites.

There is no doubt that transport of Na^+ and K^+ involves reversible changes in the conformation of the ATPase protein and that such changes may be promoted by the Na^+-stimulated phosphorylation and K^+-stimulated dephosphorylation of the pump protein. In the light of our current knowledge of transport systems in general and of the proteins of the pump in particular, it now seems unlikely that macroscopic movement of large sections of the constituent polypeptides of the pump occur as demanded by the several variations on the Lieb and Stein internal transfer model. Rather it seems that pumping may be achieved by a series of relatively small conformational changes of the protein, while the molecule as a whole does not change its disposition in the membrane. In model terms, such a system could be constituted by a pore or channel with particular binding sites for Na^+ and K^+. Such sites could be arranged transversely across the membrane, and the binding characteristics of the sites could be transiently changed by a cyclic phosphorylation/dephosphorylation of the protein. It is not unreasonable to propose that binding of an ion to one site may influence the binding characteristics of neighbouring sites, such that transfer between sites assumes a vectorial character. Although a generally accepted model has not yet emerged, it is clear that highly mobile carriers or reciprocating sub-units of polymeric transporters are less in accord with current views on the arrangement of membrane proteins than are pore or channel-forming proteins.

The basic characteristics of Na^+, K^+-ATPase from widely differing cell-types are the same, the major differences being in stoichiometry of ion movements. Exchange of one Na^+ for one K^+ will obviously not affect $\Delta\psi$, and such a pump is said to be *non-electrogenic*; the pump of the epithelial cells of frog skin is an example of a non-electrogenic Na^+, K^+-ATPase. Many of the pumps, however, are electrogenic. In most electrogenic pumps Na^+ flux exceeds K^+ flux, and in several systems a Cl^- leak counters to some extent the effect of pumping on $\Delta\psi$.

The Na^+, K^+-ATPase is predominantly associated with the plasma membrane and is concerned with maintaining the ion balance of the whole cell. Pumping of ions may also occur between compartments within cells where we might accordingly expect to find pumping proteins similar to the Na^+, K^+-ATPase. The Ca^{++}-dependent ATPase of mammalian skeletal muscle cells is one such enzyme that has been studied extensively.

The Ca^{++}-ATPase of sarcoplasmic reticulum

Skeletal muscle contraction is regulated by the level of Ca^{++} ions at the sites of interaction between the proteins of the contractile apparatus. In the relaxed muscle the calcium is sequestered within the highly vesiculated endoplasmic reticulum. Arrival of an excitatory nerve impulse at the muscle synapse causes a depolarization of the muscle plasma membrane that induces a permeability change in the sarcoplasmic reticulum such that calcium flows passively out of the vesicular system into the contractile apparatus. The ensuing contraction is terminated by removal of this calcium by the sarcoplasmic reticulum at a rate and amount consistent with relaxation of the muscle fibre. The removal of calcium involves transporting Ca^{++} back into the vesicular system against the prevailing calcium gradient, and is therefore an energy-requiring process. As in the case of the Na^+, K^+-ATPase, pumping is carried out by a membrane-bound ATPase.

The ATPase is a protein with a molecular weight of about 100 000 which spans the sarcoplasmic reticulum membrane. In the presence of Ca^{++} and ATP, a Ca_2^{++}-enzyme complex is formed which is then phosphorylated yielding a Ca_2^{++}-phosphoprotein. These stages occur at the cytoplasmic face of the membrane. Release of the two Ca^{++} ions into the lumen of the vesicles is accompanied by a dephosphorylation of the phosphoprotein, so effecting a stoichiometric coupling of Ca^{++} transport and Ca^{++}-dependent ATP hydrolysis. The transport of the Ca^{++} ions would, of course, affect $\Delta\psi$ of the sarcoplasmic reticulum, but this effect is countered by the entry of phosphate. There is still some discussion as to whether the transport of the phosphate counterion is also a function of the ATPase. Experiments in which purified ATPase is incorporated into phospholipid bilayer vesicles support this view, as phosphate is transported concurrently with Ca^{++} in such preparations. Adopting the nomenclature of Mitchell (page 187), the ATPase acts as a *symport* for Ca^{++} and phosphate.

The cyclic phosphorylation/dephosphorylation of the ATPase is reminiscent of the Na^+, K^+-ATPase, and it seems likely that a similar overall molecular mechanism may apply to the two systems. There is evidence that the binding of Ca^{++} to the protein has differential effects on the conformation of the enzyme in different regions of the molecule, and similarly the binding of ATP and the subsequent phosphorylation and dephosphorylation of the molecule may induce further changes in conformation. Thus there exists the possibility of a vectorial transfer of Ca^{++} from site to site as was postulated for the Na^+, K^+-ATPase.

The influence of the lipid components of the membrane on the Ca^{++}-ATPase has received more attention than is the case with the Na^+, K^+-ATPase. The Ca^{++}-ATPase has been isolated and incorporated into phospholipid bilayer vesicles which are seen to pump Ca^{++} with a concomitant hydrolysis of ATP. Maximum recovery of activity is seen when phosphatidylcholine and phosphatidylethanolamine are used. Work from Metcalfe and his group showed also that associated with the enzyme is a small pool of these two phospholipids that exchanges relatively slowly with other membrane phospholipids which themselves normally undergo fairly rapid lateral diffusion (chapter 6). In this way a discrete microenvironment could be maintained which particularly suits the requirements of the ATPase. The fatty acids of the phospholipids may also influence the activity of the pump. There is experimental evidence that the fatty acid composition of the membrane phospholipids changes following dietary lipid modification, and that under these conditions the calcium-pumping activity of the sarcoplasmic reticulum is changed. An enhanced Ca^{++}-ATPase activity has also been seen in cardiac muscle of hyperthyroid rats in which the levels of arachidonic and linoleic acids in the phospholipids of the sarcoplasmic reticulum showed a significant decrease.

Thus the Ca^{++}-ATPase serves to illustrate the point that, although the active transport of molecules across membranes is primarily a function of specific membrane proteins, the membrane lipids may also play an important regulatory role.

A similar Ca^{++}-ATPase is found in the plasma membrane of many cells, its function being to maintain low levels of intracellular calcium in the face of concentrations up to 10^4 times higher in the extracellular environment. The trans-plasma membrane gradients of ions maintained by the pump ATPases are widely used by cells as an energy source for transport, a topic which we shall discuss below, or for more complex membrane functions such as receptor-response coupling (see chapter 9).

Binding protein transport systems

The binding protein transport systems constitute a class of bacterial sugar and amino-acid transport processes that are dependent on ATP. The systems are characterized by having a binding protein, specific for the permeant, that is present in the periplasmic space (the area between the inner and outer membrane of the Gram-negative bacteria, chapter 1, page 7). The binding proteins are readily lost when the cells are subjected to osmotic shock. This fact accounts for a further feature of the binding-protein transport systems, which is that they cannot be successfully retained in isolated plasma-membrane vesicles.

It is generally believed that the binding proteins are involved with the transport of substrate across the outer membrane and through the periplasmic space to the plasma membrane. At this site, the binding protein may interact with a carrier that transports the substrate to the cell interior. The affinity of the binding proteins for their substrates is very high, and release of substrate would probably require an input of energy. It is presumed that it is at this point in the process that ATP is required, but there are no reports of the involvement of a specific ATPase, as is the case with the ATP-dependent ion pumps discussed above and, until the systems are better characterized, we cannot speculate further about the molecular mechanisms of the process.

Mitchell's chemiosmotic theory

We discussed earlier the concept that membrane-bound enzymes might assume a vectorial character and thereby constitute a group translocation transport system. Mitchell also suggested that there may be membrane proteins that could achieve the same transport effects without the necessity for chemical modification of the transported molecule, and to such proteins he gave the name *porters*.

The fundamental conceptual difference between Mitchell's ideas and those of other workers was the nature of the energy source for the transport processes. Mitchell was almost alone in abandoning the dogma of established biochemistry that metabolic energy is always conserved in the form of high-energy chemical intermediates, and in suggesting that in many cases energy is stored as a trans-membrane electrochemical gradient of protons. Despite considerable scepticism, the hypothesis progressed to the point where it became a paradigm for transport systems and energy-transducing membranes, and the *chemiosmotic theory* received its ultimate vindication with the award of the Nobel prize for Chemistry to Mitchell in 1978.

Let us look in some detail at the possible role of a transmembrane proton gradient as a source of energy. A prerequisite for the system is that the membrane is impermeable to protons, and in general this is seen to be so. Next, there should be mechanisms that actively pump protons across the membrane. These mechanisms are dealt with in detail in the next chapter; for the moment we will accept that such pumps are a common feature of a variety of different membrane types. Thus a gradient of protons can be established and can be measured directly as a pH difference between the two media separated by the membrane. The transport of protons in the absence of movement of other ions will affect $\Delta\psi$. Therefore, as we saw earlier in our considerations of ion concentration gradients, the chemical concentration gradient of protons ΔpH and $\Delta\psi$ must be considered together, and their relative contributions to the final *electrochemical concentration gradient* of protons $\Delta\tilde{\mu}_{H^+}$ is given by

$$\Delta\tilde{\mu}_{H^+} = F\Delta\psi - 2\cdot3\ RT\Delta pH$$

where F = the Faraday;
$\quad R$ = the gas constant;
$\quad T$ = temperature in K.

Mitchell has rearranged this equation and expressed it in a form reminiscent of the simplified Nernst equation that we discussed on page 179 such that

$$\Delta p = \Delta\psi - Z\Delta pH$$

where Z is a combination of constants equal to approximately 59 mV/pH unit at 37° C, and Δp is the *proton motive force* or pmf. Thus Δp is a measure of the potential energy of the proton gradient. Many membranes utilize Δp to perform transport work. How is the coupling achieved? In many systems the transport system has the properties of what, in Mitchell's terminology, is a *proton symport*. This means that the porter for the transported solute is also a porter for protons; so long as $\Delta\tilde{\mu}_{H^+}$ remains greater than the concentration gradient of the solute, transport will continue (figure 8.5).

There is abundant evidence that the concept of a symport driven by a transmembrane electrochemical gradient is also valid for non-proton-dependent transport mechanisms. Sodium particularly is seen to have an

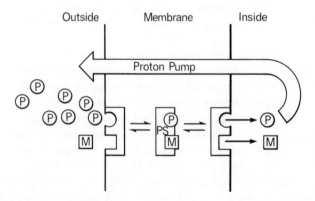

Figure 8.5. The proton symport: proton symport PS combines with protons P and metabolite M at the outer face of the membrane. The protons run down their concentration gradient and power the active transport of M. The symport is shown here as a mobile carrier but it is emphasized that this is purely schematic (see page 193).

analogous role to the proton in the transport of amino acids and sugars in both prokaryotes and in eukaryotic cells. In the remaining part of this chapter we shall look at examples of chemiosmotic transport systems, and attempt to describe their workings in molecular terms.

Lactose transport in E. coli

Deere and his co-workers in 1939 showed that disrupted cells of *E. coli* hydrolyze β-galactosides more rapidly than do intact cells, suggesting that the cell membrane is not freely permeable to the sugars. Later work showed that β-galactosides, lactose particularly, are transported across the membrane and appear in the cell in a chemically unmodified form. The characteristics of the system were established by Cohen and Monod and their co-workers in the Pasteur Institute, and it has since served as a model for investigations into the molecular processes of membrane transport in many other systems. The lactose transport system comprises a single protein, the product of the *lacY* gene which is a component of the lactose operon. The lactose transport protein is inducible and, as with all bacterial systems, the relative ease with which genetic variants of the 'parent' organism may be obtained has facilitated the very extensive characterization of the regulatory aspects of the system. The transport protein itself is not, however, well defined.

In 1965 in a now classical series of experiments, Fox and Kennedy covalently labelled a membrane protein that they suggested was the lactose carrier. It was known that the transport system is sensitive to the sulphhydryl-modifying reagent, N-ethylmaleimide, from which it can be protected by β,β'-thiodigalactoside. Fox and Kennedy reasoned that treatment of *E. coli* with N-ethylmaleimide in the presence of β,β' thiodigalactoside would result in covalent modification of all membrane SH groups except that of the lactose transporter to which the β,β' thiodigalactoside was bound. Removal of the galactoside and subsequent treatment with radiolabelled N-ethylmaleimide would then specifically label the carrier. The experiments resulted in the isolation of a detergent-solubilized membrane protein to which Fox and Kennedy gave the name M protein. Further work showed that M protein was the product of the *lacY* gene. The protein has not yet been completely purified, but it seems to be a small (30 000 molecular weight) hydrophobic protein that can only be maintained in aqueous solution by detergents. There are reports that addition of a solubilized preparation of M protein to membrane vesicles prepared from carrier-deficient *lacY* mutants confers energy-dependent lactose uptake on the vesicles, thus providing further evidence for the transporter role of the protein.

There is abundant evidence that the lactose transporter is a proton symport. Early experiments of West and Mitchell showed that uptake of β-galactosides by *E. coli* resulted in alkalinization of the medium. This change in medium pH did not occur with cells genetically deficient in the carrier, or in normal cells treated with N-ethylmaleimide or p-chloromercuribenzoate. It was also shown that galactose transport is an *electrogenic process*, i.e. passage of an uncharged sugar plus a proton across the cell membrane changes $\Delta\psi$. By using ^{14}C lactose and measuring proton movements, Mitchell and West were able to show a stoichiometry of 1g ion H^+/mole of lactose.

We must conclude that protonation of the carrier, the M protein, changes its affinity for lactose. Thus, at the outer face of the membrane, the protonated form of the carrier, having a high affinity for lactose, predominates; at the inner face, the proton dissociates and the affinity for lactose decreases, leading to release of the sugar. The asymmetric distribution of the two forms of the carrier is maintained by $\Delta\tilde{\mu}_{H^+}$.

There are other sugar transport systems known to operate in *E. coli* and other micro-organisms but none is as well characterized as the lactose system. In several systems specific binding proteins have been isolated and there is evidence that in some cases coupling to $\Delta\tilde{\mu}_{H^+}$ provides the driving force for active accumulation.

Sodium-dependent transport systems

A specific dependence on sodium ions has been observed for the transport of amino acids and sugars into a wide variety of both prokaryotic and eukaryotic cell types. Christensen first reported that the uptake of glycine or alanine into duck erythrocytes was absolutely dependent on Na^+ in the medium, and that replacement of Na^+ by K^+ resulted in a complete inhibition of transport. Since these early experiments, there has emerged definitive evidence that in cell types ranging from bacteria through goldfish epithelia and bullfrog intestine to human intestine and brain, the active uptake of many amino acids and sugars is a sodium-dependent process and, furthermore, that the energy for transport derives from the sodium concentration gradient. There is also evidence for a similar sodium-dependent transport of the anions sulphate, phosphate and acetate in some mammalian tissues.

Recalling our discussion of proton gradients (page 186), the electrochemical concentration gradient of Na^+ ($\Delta\tilde{\mu}_{Na^+}$ is made up from two components, $\Delta\psi$ and ΔNa^+ (the transmembrane concentration gradient). The membrane utilizes the potential energy of $\Delta\tilde{\mu}_{Na^+}$ in the sodium-dependent transport systems. Thus on association with Na^+ the sugar and amino-acid transporting systems become positively charged symports and respond to both components of $\Delta\tilde{\mu}_{Na^+}$. The electroneutral symports, such as those carrying Na^+ together with phosphate or sulphate, are driven only by the ΔNa^+ component of $\Delta\tilde{\mu}_{Na^+}$. In the case of the bacterium *Halobacterium halobium* an Na^+/H^+ electroneutral antiport has been described; again this is driven by ΔNa^+.

A recent series of experiments, in which the sodium-linked melibiose transport system of the bacterium *Salmonella typhimurium* has been examined, provides evidence for distinct contributions from ΔNa^+ and from $\Delta\psi$ to the overall transport mechanism. The porter is perturbed directly by $\Delta\psi$, which alters its interaction with Na^+ and allows the cation to effect an increase in binding affinity. If we recall the discussion on the colicins (page 173), we see here a further example of a change in electric field inducing a conformational change in a membrane protein, a phenomenon to which we shall return and consider in detail in chapter 9. In the case of the melibiose porter, an increased $\Delta\psi$ increases the affinity of an Na^+ binding site. The porter is affected in a quite different way by ΔNa^+ which independently of $\Delta\psi$ increases the number of melibiose binding sites, and it is suggested that Na^+ binding precedes solute binding and translocation.

The functioning of the porters will, of course, dissipate $\Delta\mu_{Na^+}$ and active transport of solute would rapidly cease unless the sodium gradient is maintained. In eukaryotic cells the Na^+, K^+-ATPase is the major contributor to $\Delta\tilde{\mu}_{Na^+}$ and in cells with a particularly high transport activity (e.g. the kidney tubule cells which must recover great quantities of glucose and amino acids from the glomerular filtrate) up to 80% of the cells' production of ATP is utilized by the Na^+, K^+-ATPase. In many micro-organisms, the energy of the sodium gradient derives from the proton gradient, i.e. $\Delta\tilde{\mu}_{H^+}$ generates $\Delta\tilde{\mu}_{Na^+}$ via an Na^+/H^+ antiport system. However, as in the eukaryotic cell, there is still a requirement for ATP, as $\Delta\tilde{\mu}_{H^+}$ is generated and maintained by a plasma membrane-bound 'coupling'- or H^+-ATPase which we shall discuss in chapter 9.

A variation of the sodium-linked transport system in which the symport and the Na^+, K^+-ATPase reside in different membranes of the same cell is found in the mucosal cells of the small intestine. These cells have one face, the mucosal surface, lining the lumen of the intestine, while the opposite serosal face borders the blood stream. The sodium concentration in the lumen is higher than that in the mucosal cell, and a slow leakage of Na^+ occurs across the mucosal membrane. The Na^+, K^+-ATPase, which resides in the serosal membrane, pumps the Na^+ out into the blood. Thus the overall effect is a transfer of Na^+ from the lumen to the blood, and it is seen that in the resting state the lumen is about 3 millivolts negative with respect to the blood. The mucosal membrane contains a glucose-Na^+ symport that is powered by the Na^+ gradient, and carries glucose into the cells from the intestinal lumen. During active glucose uptake the Na^+ gradient across the mucosal membrane is maintained by increased pumping activity across the opposite serosal membrane, and this is reflected in the potential difference between the lumen and the blood which rises to 10 millivolts.

Thus in this system the transport of one molecular species across a membrane is effectively coupled to the active transport activity of a quite separate membrane. The serosal membrane also has a glucose transport system, but this has a sodium-independent facilitated diffusion mechanism

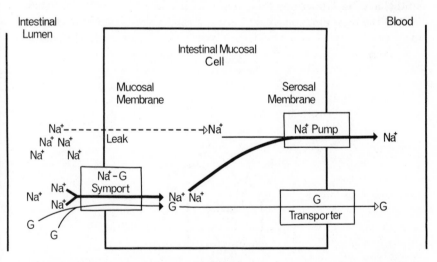

Figure 8.6. The Na^+-linked transport of glucose in intestinal mucosal cells: in the resting state Na^+ leaks across the mucosal membrane into the cell and is pumped out across the serosal membrane into the blood. The Na^+ gradient provides the energy for the uptake of glucose G which enters the cell on an Na^+-G symport. The Na^+ gradient across the mucosal membrane is maintained by pumping the Na^+ from the cell interior into the blood. Glucose enters the blood via a facilitated-diffusion transporter located in the serosal membrane.

and aids the passage of glucose from the high-concentration area within the mucosal cell to the lower-concentration area of the blood (see figure 8.6).

The epithelial cells of the small intestine provide an example of a quite different mode of coupling of transport to sodium ions. There is an active reabsorption of water from the lumen of the small intestine, such that about 95 % of the water is recovered. The movement of this amount of water across the intestinal epithelium could not occur simply as a result of the osmotic content of the epithelial cells, but is coupled to pumping of Na^+ ions. The epithelial cells are sealed by tight junctions (see chapter 3, page 27), basolateral to which the cells tend to separate, forming restricted extracellular channels. There is an active pumping of Na^+ into the channels immediately below the tight junctions, and the resulting $\Delta\psi$ effects the electrophoresis of Cl^- into the space, thus generating a local area of high salt concentration. Water is drawn into this space by osmosis from the intestinal lumen via two distinct routes. It is suggested that as much as 50 % of the water entering the space crosses the tight junction directly, and the remainder is drawn out of the epithelial cells. Under normal conditions, the reflection coefficients (page 167) of the tight junction to Na^+ and Cl^- are much higher than those of the membranes of the blood capillaries which constitute the serosal

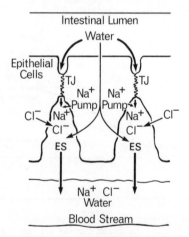

Figure 8.7. Water reabsorption in the small intestine: the pumping of Na^+ into the extracellular space ES, followed by electrophoresis of Cl^- generates a high salt concentration. Water is drawn into ES by osmosis through the tight junction TJ and from the epithelial cells. The back-flow of salt is not possible under normal conditions, and there is a net flow of salt solution towards the base of ES and into the blood capillaries.

border of the extracellular spaces. Thus salt cannot cross the junction into the lumen, and there is a flow of salt solution towards the base of the channels, into the serosal extracellular space and then into the blood capillaries. The system is summarized in figure 8.7.

The overall effect is to transfer water from the intestinal lumen to the blood, but the transfer will continue only so long as the local high salt concentration is maintained by active Na^+ pumping. Under some circumstances the permeability of the tight junctions may change, thus cholera toxin (see chapters 7, page 154 and 9, page 232) induces a marked reduction in the reflection coefficients of the junction to Na^+ and Cl^- such that a major leakage of these two ions into the intestinal lumen occurs; water uptake therefore ceases. The cholera victim suffers massive salt and water loss, leading to a degree of tissue dehydration that can rapidly prove fatal.

The active transport of water by this system is clearly achieved by a quite different mechanism from those of the various sodium symport processes. However, there is a common basic pattern, and that is the conversion of the metabolic energy of the cell (ATP) into an osmotic energy-source ($\Delta\tilde{\mu}_{Na^+}$) and its utilization for transport work.

We have as yet no clear idea of the molecular nature of the various porter systems that we have considered in this section. We can presume that the symports are membrane proteins with specific binding

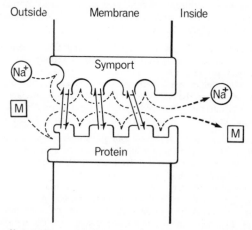

Figure 8.8. The sodium symport: binding of Na^+ to sites at the outside face of a channel formed by a trans-membrane protein induces a change in the affinity of a site for metabolite M. Sequential binding of Na^+ to sites along the protein induces changes in the metabolite binding sites such that a vectorial sequential binding occurs.

sites for both Na^+ and the permeant. A diffusing carrier as depicted in figure 8.5 provides a schematic model for the symports but, as stated previously (page 174), current thinking on transporters is more in favour of a relatively stationary protein with perhaps a series of binding sites spanning the membrane. As considered in our discussion of the Na^+ K^+-ATPase, it is possible that minor conformational changes in adjacent binding sites for permeant induced by Na^+ binding and by transfer along a series of Na^+-dependent sites could result in a vectorial sequential binding of the permeant as depicted in figure 8.8.

In this chapter we have restricted our discussion to mechanisms for transporting permeants across membranes. To some extent we have pre-empted a consideration of phenomena in which a transport component is but a part of a more complex membrane function such as, for example, the $\Delta\psi$-dependent formation of transmembrane channels. Some of these activities we can begin to understand in molecular terms, and we shall consider a selection of them in the next chapter.

SUMMARY

1. All membranes show great selectivity in their permeability to solutes. Small solute molecules having a radius less than 0·5 nm may diffuse through water-filled channels that traverse most membranes.

2. The transport of most permeants is facilitated by specific membrane proteins termed carriers or transporters. These proteins probably provide the permeant molecule with an environment that replaces the hydration shell.

3. Facilitated diffusion carriers provide an energy-independent method for the specific transfer of permeants across membranes. Transport continues only so long as a concentration gradient of permeant is maintained. Current opinion favours molecular mechanisms in which the carrier constitutes a specific trans-membrane pore or channel. An alternative view is that the carrier ferries permeant molecules across the membrane. A diverse group of natural compounds grouped together as the ionophoric antibiotics provide molecular models for both interpretations of carrier function.

4. Active transport systems use a variety of energy sources to power the uptake of solutes against a concentration gradient. Group translocation systems involve the chemical modification of the permeant as it is transported. Such systems have been well characterized in micro-organisms but are less well established in higher organisms. ATP provides the energy for transporting some permeants and, in the case of the active pumping of Na^+, K^+ and Ca^{++}, the transporter protein has ATPase activity. Many bacterial transport systems are also dependent on ATP, but their mechanisms have not been elucidated. Mitchell's chemiosmotic hypothesis provides for the coupling of the potential energy of trans-membrane electrochemical gradients to specific transport systems. Proton gradients particularly are widely used in both prokaryotic and eukaryotic cells. Analogous sodium-dependent transport systems are also found in a variety of cell types but are best characterized in mammalian cell membranes.

CHAPTER NINE

COMPLEX MEMBRANE-MEDIATED PROCESSES

CHAPTER TWO OUTLINED THE DIVERSE WAYS IN WHICH MEMBRANES influence the activities of cells. Metabolism occurs within discrete compartments between which communication can occur only via the boundary membranes. In addition to their role as selectively permeable barriers, some aspects of which we have discussed in chapter 8, the membranes may also participate directly in the events occurring within the metabolic compartment. Thus many of the enzymes of a metabolic sequence may be membrane-bound, and as such their activities can be regulated by the membrane.

One of the most intensively studied membrane-bound enzyme systems is the proton-translocating or coupling ATPase to which we have already made brief reference (chapter 8, page 190).

Energy transduction and ATP synthesis

In chapter 8 we discussed the chemiosmotic theory of Mitchell and saw how the energy of a transmembrane proton gradient could be used to do transport work. The coupling ATPases use this same energy source to synthesize ATP, and in this section we shall examine the mechanisms by which this is achieved. There are three major groups of energy transducing membranes: the thylakoid membranes of chloroplasts, the plasma membrane of the prokaryotes, and the inner mitochondrial membrane. The ATPases from these three sources are similar in many respects, and the membranes have closely similar mechanisms for generating the transmembrane proton gradient. The membranes do differ, however, in the nature of the primary energy source—this being light in the case of the thylakoid membrane, whereas oxidative metabolism provides the energy for the prokaryote plasma membrane and for the inner mitochondrial membrane.

195

The coupling ATPase

The ATPases from the three membrane sources are remarkably similar to each other but show few common features with the ion pumping ATPases discussed in chapter 8 (page 178). Indeed, so alike are the three enzymes that it is suggested that they have evolved from a common ancestral prokaryotic protein, a view in accord with the endosymbiont theory of the origins of mitochondria and chloroplasts.

The ATPase is made up from two distinct classes of subunit. The catalytic unit is referred to as the F_1-ATPase or more usually, simply as F_1. This is a peripheral or extrinsic protein (page 90, chapter 5) and is soluble in water once detached from the second type of sub-unit F_0, which is an integral protein. The F_0 fragment is believed to form the pore through which protons flow across the coupling or transducing membrane. F_1 corresponds to the knobs and stalks which project from the matrix side of the inner mitochondrial membrane and which are described in chapter 3 (page 23); similar particles may be seen on thylakoid membranes and on the cytoplasmic face of bacterial membranes.

The two parts of the ATPase can be easily separated. They are connected through both ionic and hydrophobic interactions, and it is thought that Mg^{++} may act as a bridge between anionic sites on the two proteins. Both F_1 and F_0 have been purified to homogeneity from several different sources. F_1 is a large molecule of total molecular weight about 380 000, and comprises five distinct sub-units ($\alpha \sim 58\,000$; $\beta \sim 52\,000$; γ 32 000; δ 17 000 and ε 12 000). The sub-units from F_1 of bacteria, chloroplasts and mitochondria show similar size distribution, and sub-units of the same size have been assigned similar functions in the ATPase from all three sources. These activities are:

α and β components of the ATPase catalytic site, together showing ATPase activity in the absence of γ, δ and ε.

γ necessary for reassembly of α and β after cold inactivation (see page 13); together with ε may inhibit ATPase.

δ necessary for association of F_1 and F_0 and thought to be the 'stalk'.

ε inhibits ATPase of F_1, possibly has a regulatory role *in vivo* (particularly good evidence for this in mitochondria).

The F_0 component is not quite so well characterized. It is a multi-subunit protein of molecular weight 150 000, comprising four very hydrophobic peptides which are thought to span the membrane, and a further peptide which is hydrophilic and is exposed at the membrane surface. The protein shows no enzymic activity but confers a specific proton permeability on the membrane. Experiments with *E. coli* mutants provide direct evidence for such a role. The mutants lack ATPase activity owing to a defective F_1 but F_0 is not affected: the permeability of the plasma membrane of these mutants to protons is greatly enhanced. Treatment of the cells with dicyclohexylcarbodiimide (DCCD) seals the proton leak. DCCD is well known to inhibit the ATPase in normal membrane systems. A single small proteolipid has been identified as the site of action of DCCD on F_0 in both *E. coli* and mitochondria, and it is attractive to postulate that this particular sub-unit of F_0 may be a transmembrane proton channel.

The ATPase and $\Delta\tilde{\mu}_{H^+}$

The reaction catalyzed by the ATPase is usually written

$$ADP + P_i \rightleftharpoons ATP \tag{i}$$

or more fully:

$$ADP^{3-} + H_2PO_4^- \rightleftharpoons ATP^{4-} + H_2O \tag{ii}$$

The equilibrium for the reaction lies far to the left, i.e. ATP hydrolysis releases considerable energy, and equally a large input of energy is required to achieve ATP synthesis. The ATPase seems also to act as an obligatory proton pump, extruding two protons through F_0 as one molecule of ATP is cleaved at F_1. So we can modify (ii) to

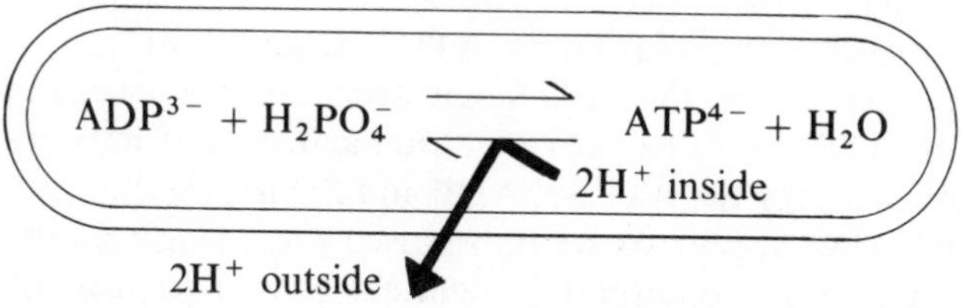

$$ADP^{3-} + H_2PO_4^- \quad \longrightarrow \quad ATP^{4-} + H_2O \tag{iii}$$

2H⁺ inside

2H⁺ outside

Thus the ATPase can act as an electrogenic proton pump and establish a $\Delta\tilde{\mu}_{H^+}$. Such an activity is seen in the anaerobic bacteria in which an ATP-generated proton gradient provides an energy source for solute uptake (see page 190). However, let us suppose that some other proton-pumping activity of the ATPase-containing membrane establishes an opposing $\Delta\tilde{\mu}_{H^+}$ that exceeds that generated by ATP hydrolysis. Protons will now flow back through F_0 and drive the reaction of the ATPase in the direction of synthesis:

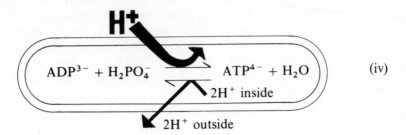

$$ADP^{3-} + H_2PO_4^- \rightleftharpoons ATP^{4-} + H_2O \qquad \text{(iv)}$$

Direct evidence for the activities proposed in (iii) and (iv) is provided by experiments in which purified mitochondrial ATPase was incorporated into liposomes which then pumped protons on the addition of ATP. Addition of *bacteriorhodopsin* resulted in ATP synthesis when the liposomes were illuminated. As we see in a later section (page 207) bacteriorhodopsin acts as a light-stimulated proton pump generating a proton gradient which, in this case, was sufficient to reverse the ATP hydrolysis. These experiments provide some of the most convincing evidence for the chemiosmotic interpretation of the mode of synthesis of ATP by the coupling ATPase.

ADP/ATP transport

The arrangement of the coupling ATPase in the inner mitochondrial membrane means that ATP is synthesized within the mitochondrial matrix. However, as we well know, ATP is required for a multitude of metabolic reactions in the cell. What then are the mechanisms for transporting ATP out of the mitochondrion and, equally important, how are the substrates for the ATPase, ADP and P_i brought into the matrix?

ATP and ADP appear to be transported on a single antiport. The molecular nature of the antiport system has been investigated, and the Klingenberg group particularly has taken tentative steps towards identifying it with a particular protein. Their studies have been aided by the availability of two very specific inhibitors, atractylate and bongkrekate. The inner membrane is impermeable to atractylate but is permeable to bongkrekate. Thus atractylate can interact with the transporter from the intermembrane space only, whereas bongkrekate can approach the transporter from either side. These studies have led the Klingenberg group to conclude that the transporter has only a single reorienting site, i.e. a site on the carrier picks up ADP at the intermembrane space surface and transports it into the matrix, ATP is then transported out of the matrix on the return journey. The matrix is negative with respect to the intramembrane space, and it is suggested that the ATP/ADP antiport is

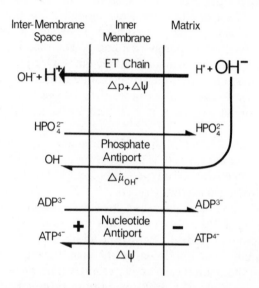

Figure 9.1. Transport systems of the inner mitochondrial membrane associated with ATP synthesis: the electron transport chain, E.T. Chain, generates a trans-membrane proton gradient (see page 200). Phosphate enters on an OH^--phosphate antiport driven by $\Delta\tilde{\mu}_{OH^+}$. ATP and ADP are transported by an antiport that is driven by $\Delta\psi$.

driven by the $\Delta\psi$ component of $\Delta\tilde{\mu}_{H^+}$ as ATP has an additional negative charge compared to ADP. In addition to ADP, ATP synthesis requires also inorganic phosphate. This is thought to enter the matrix via an electroneutral antiport shared with hydroxyl ions, utilizing the energy of the $\Delta\tilde{\mu}_{OH^-}$ that is generated by proton pumping. These systems are summarized in figure 9.1.

The inner membrane also possesses a number of specific transport systems for substrates and constituents of the TCA cycle. However, apart from being identified in terms of the permeant, e.g. a malate/citrate or malate/αketoglutarate antiport, the carriers have not been characterized.

The generation of the proton gradient

It is now generally accepted that organisms convert the energy of absorbed light or the energy released by oxidation of substrates into chemical energy as ATP via a proton gradient, the processes being respectively *photosynthetic phosphorylation* and *oxidative phosphorylation.*

Before we move on to consider the details of the process, we should consider briefly the pre-chemiosmotic theory view of how such coupling was achieved. The early theories were modelled on the general biochemical mechanism of substrate-linked phosphorylation, in which energy is conserved in an 'energy-rich bond'. The theories predicted the formation of high-energy chemical intermediates which were common to the mitochondrial and chloroplast electron transport chains and ATP synthesis. Intensive research over more than 20 years failed to produce any evidence for the existence of such intermediates, and the theory is now defunct.

The generation of the proton gradients in the three coupling membranes that we are considering is achieved by anisotropically orientated electron transport chains. Two types of molecules constitute such chains: these are the hydrogen carriers (transporting $2H^+ + 2e^-$) which undergo sequential reduction and oxidation by the acquisition and loss of hydrogen atoms, and the electron carriers (transporting $2e^-$) which are similarly reduced and oxidized, but by the addition and loss of electrons. The carriers are arranged in 'loops' in the membrane so that passage of electrons down the chain (i.e. to an increasingly positive redox potential) results in the build-up of a trans-membrane proton gradient. In the mitochondrial inner membrane and the plasma membrane of aerobic bacteria, the overall flow of electrons is from NADH (NADH = $NAD^+ + H^+ + 2e^-$) to oxygen ($\frac{1}{2}O_2 + 2H^+ + 2e^- = H_2O$) giving water. This is an exergonic process and the energy released is made available to generate a proton gradient. Alternatively $FADH_2$ may be the initial electron donor. In green plants and photosynthetic bacteria, electrons are transferred in the reverse direction from water to $NADP^+$, an endergonic process, and the energy required to effect this transfer and

Figure 9.2. Proton and electron-carrying loops in energy-transducing membranes. The upper part of the diagram summarizes the flow of electrons and of protons in the inner mitochondrial membrane and in the plasma membrane of the aerobic bacteria. The overall flow of electrons is from NADH to oxygen; NADH has a higher (i.e. more negative) redox potential than does oxygen, and the passage of electrons down the chain of loop carriers releases energy which is used to generate the proton gradient. The flow of electrons through the loop carriers in a photosynthetic membrane, shown in the lower diagram, is in the reverse

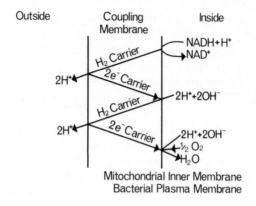

Mitochondrial Inner Membrane
Bacterial Plasma Membrane

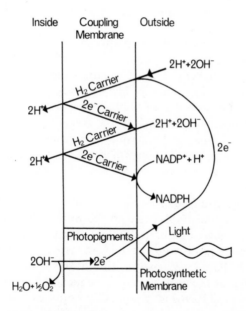

direction: from water to $NADP^+$. Water has a lower (i.e. more positive) redox potential than $NADP^+$, thus an input of energy is required. This is provided by the photopigments which on absorption of light energy are promoted to an excited state. Electrons from the activated pigments are fed into the loop carrier system and the electron-deficiency of the pigments is made good by the donation of electrons from the products of the ionization of water. It should be noted that the diagrams are designed only to illustrate general concepts and do not describe particular membranes in detail (see page 203).

simultaneously to establish a proton gradient is obtained from light excitation of photopigments. The concept of the loop carrier electron transport chains is summarized in figure 9.2. In the higher plants the transduction process occurs at the thylakoid membranes of the chloroplast, whereas in the photosynthetic bacteria specialized invaginations of the plasma membrane form the photosynthetic apparatus (see chapter 1, page 8).

Two types of light absorption system are found in the thylakoid membrane and these are termed photosystems I and II. Photosystem II centres around a protein-bound form of chlorophyll called P_{682}. This complex is excited either directly by light absorption, or by resonance energy transfer from associated carotenoids and chlorophylls. There is evidence that the latter route is the major one. Activated electrons from P_{682} pass through a series of loop carriers as outlined above. The process clearly leaves P_{682} deficient in electrons, and this requirement is met from the products of the ionization of water (see figure 9.2) but the details of the reaction are not yet clear.

The final carrier associated with photosystem II is plastocyanin which can donate the electrons to photosystem I. Photosystem I raises the electrons to a new higher (i.e. more negative) redox potential as a result of a second input of light energy. The light energy is transduced by an analogous mechanism to that of photosystem II but involves a different type of protein-bound chlorophyll (P_{700}). Again the electrons pass down a series of carrier loops and build up a transmembrane proton gradient. The final carrier is a flavoprotein which utilizes the electrons to reduce $NADP^+$; the proton also required by NADP is provided by the ionization of water.

The carriers are arranged in such a way that protons are taken up from the external phase and transported to the intra-thylakoid space (figure 9.2). It has been shown that under illumination the internal phase of the thylakoids becomes positive and acidic with respect to the external phase. The coupling ATPase is attached to the thylakoid membrane from the outer side, and thus the protons can subsequently flow out of the thylakoid through the ATPase and power ATP synthesis (see figure 9.3).

There is a wealth of experimental evidence that supports the general scheme for photosynthetic phosphorylation outlined above, but it should not be assumed that we have a complete understanding of the molecular events that achieve the overall coupling of light energy to ATP synthesis. The identity of several of the components of the loop carriers is still unknown. Particularly, we have little knowledge of the components that

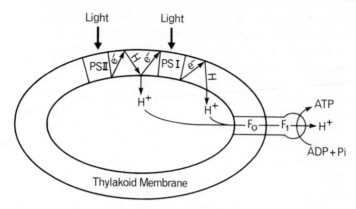

Figure 9.3. Photosynthetic phosphorylation.

accept and conduct the electrons that are photochemically ejected from the photosystems to the first of the loop carriers, plastoquinone (photosystem II) or ferredoxin (photosystem I).

There is also some controversy over the stoichiometry of the proton-electron transport process. Experiments made under intermittent illumination are at variance with those performed under continuous light, although a value of $2H^+/e^-$ has been reported by some groups using both types of illumination. The stoichiometry of protons translocated to ATP molecules synthesized also varies between research groups, the values ranging from two to four. However, although much work remains to be done before such details are clarified, there is now little dissent from the view that, in general terms, photosynthetic phosphorylation proceeds in accordance with Mitchell's chemiosmotic hypothesis.

The components of the inner mitochondrial membrane and the plasma membranes of aerobic bacteria that constitute the respiratory or electron-transport chain are, in general, better characterized than their photosynthetic counterparts. In the case of the mitochondrial systems, the components of the chain have all been identified, and the positions of many of these within the membrane are known, although in some cases there is still disagreement about their precise sequence. The spatial relationship of the components of the inner membrane has been determined largely by the use of specific redox reagents of limited permeability, coupled with vesicular preparations of the inner mitochondrial membrane that can be prepared with either the normal sidedness or in an inside-out state. Thus, ferricyanide does not penetrate the membrane but can be shown to interact with cytochromes c and c_1. This holds true in normal-sided vesicles but is not seen in inside-out vesicles, indicating that c and c_1 must be located at the membrane's outer surface. Azide, on the other hand, which inhibits respiration at the stage of cytochrome c_3, has a much more pronounced effect on inside-out vesicles than on normal-sided preparations, suggesting that c_3 is exposed at the matrix side of the inner membrane. Antibodies to F_1 have no effect on intact mitochondria or on normal-sided vesicles, but inhibit ATP synthesis in inside-out vesicle preparations, a finding consistent with the view that the ATPase projects into the matrix.

The localization of some components is still not known. Thus the b-cytochromes and the rather ill-defined iron-sulphur proteins are presumed to reside in the hydrophobic interior of the membrane, but are not known to be associated with either face. Figure 9.4 shows the currently accepted orientation of the chain in the membrane.

As discussed previously (page 200) the components of the chain are thought to form a system of loop carriers for protons and electrons. However, it is not yet possible to assign specific hydrogen and electron

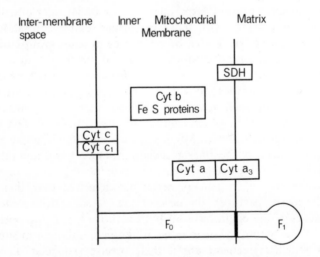

Figure 9.4. Orientation of the electron transport chain within the inner mitochondrial membrane: succinate dehydrogenase, SDH, is located at the inner face of the membrane. The F_1 portion of the ATPase and cytochrome a_3 are also located at this site. Cytochromes c and c_1 are accessible from the inter-membrane space and are thought to be at the membrane surface. The b cytochromes and the iron-sulphur proteins, FeS proteins, are located within the hydrophobic region of the membrane.

carrier functions to each component of the chain. There is an alternative view that the individual chain constituents can act as proton pumps. However, although as we shall see in the next section (page 207), the bacterial protein bacteriorhodopsin can act as a direct proton pump, there is so far little evidence that the single constituents of the mitochondrial electron transfer chain can carry out such an activity. What evidence there is comes from experiments in which cytochrome oxidase alone has been shown to translocate protons across the membrane of single bilayer vesicles.

Although there is still considerable discussion about the details of the mechanisms by which a proton gradient across the inner mitochondrial membrane is produced, it is generally accepted that ATP synthesis is powered by such a gradient. The gradient is such that protons run down their concentration gradient into the matrix and energize the inward-facing coupling ATPase (figure 9.5).

It can be appreciated that the systems that we have examined demand that the coupling membrane itself, i.e. the membrane in which the ATPase resides, is impermeable to protons. This can be verified experimentally. It is also apparent that should the membrane become leaky to protons, then the $\Delta\tilde{\mu}_{H^+}$ will be dissipated, and work that is dependent on the proton gradient will cease. This is the explanation for the action

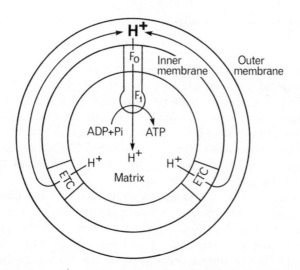

Figure 9.5. The coupling ATPase in the inner mitochondrial membrane.

of *uncouplers* of oxidative phosphorylation. These are compounds which prevent the phosphorylation of ADP and are capable of causing a parallel stimulation of respiration. The uncouplers are a molecularly-diverse group of compounds, but Mitchell pointed out in the early 1960s that each is a lipophilic weak organic acid which can readily diffuse across the membrane. In the presence of $\Delta\tilde{\mu}_{H^+}$ the uncoupler becomes

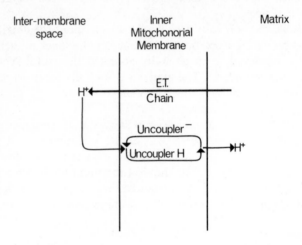

Figure 9.6. Uncouplers dissipate the proton gradient by acting as transporters for protons.

protonated at the face of the membrane in contact with the higher proton concentration, and dissociates at the opposite face, releasing the proton into the lower concentration environment. Thus the uncouplers act as ionophores (see page 171) for protons and dissipate $\Delta\tilde{\mu}_{H^+}$. ATP synthesis therefore ceases. It is suggested that the stimulation of respiration occurs by a mass action effect; the build-up of protons is relieved by the uncoupler, thus allowing electron flow and proton extrusion to proceed unhindered (figure 9.6).

The energy-transducing membranes that we have considered so far are highly complex structures, and we cannot claim to have a complete understanding of the molecular bases of their functions. There is, however, a membrane system for which such a goal could be said to be in sight: this is the purple membrane of the bacterium *Halobacterium halobium*.

The purple membrane of *Halobacterium halobium*

Halobacterium halobium is one of a group of extreme halophiles—the *Halobacteria*, whose natural habitats are salt flats and stagnant pools at the edge of tropical seas. Under such conditions, salt concentrations close to saturation are maintained, temperature and solar radiation are high, and the oxygen content is low. To cope with this environment, *H. halobium* has evolved a fascinating plasma membrane which offers what is currently a unique attraction to the researcher—it contains only a single protein which can be readily isolated and is well characterized.

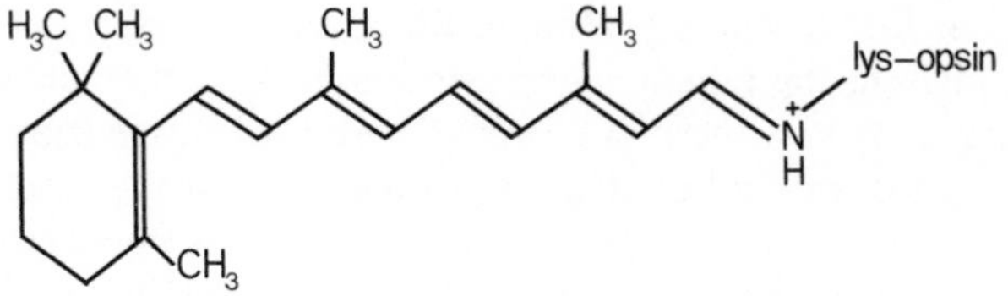

Figure 9.7. Bacteriorhodopsin, showing the chromophore *retinal* in the all-*trans* configuration.

When grown in the light in at least 12% salt and at low oxygen tension, *H. halobium* synthesizes patches of plasma membrane containing a purple pigment. In 1967 Walter Stoeckenius and his collaborators isolated for the first time these purple fragments and went on to show that they contain a single species of polypeptide with molecular weight 26 000, and that the purple colour is caused by the presence of retinal which is attached to the amino group of a lysine residue on the polypeptide (figure 9.7). As retinal was, until that time, known to occur only in the visual pigment rhodopsin (see page 36), the protein-retinal complex was called *bacteriorhodopsin*. Unlike rhodopsin, bacteriorhodopsin seemed to be arranged as an exact two-dimensional crystal in a rigid planar hexagonal lattice throughout the purple membrane. In 1973 Oesterhelt and Stoeckenius proposed a remarkable function for the purple membrane; they suggested that bacteriorhodopsin could act as a light-driven proton pump which would create a proton gradient across the plasma membrane, and that the cell used this to drive ATP synthesis. We now know that this is so, and we are approaching an understanding of the molecular nature of the pump mechanism.

It has been confirmed that bacteriorhodopsin comprises a protein of molecular weight 26 000 to which is bound a single retinal moiety. Bacteriorhodopsin shows a broad absorption peak at about 570 nm which accounts for the deep purple colour of the membrane. In the light, the retinal is present as the all-*trans* isomer (as in figure 9.7) and it is thought that this is the form which predominates under physiological conditions. A flash of visible light promotes a cyclic sequence of absorbance changes in which five distinct photo-intermediates are observed. Concomitant with this reaction is an asymmetric loss and re-uptake of a proton, leading to a net movement of protons across the membrane. Flash spectroscopy and resonance Raman spectroscopy have revealed the points in the cycle at which proton loss and uptake occurs (figure 9.8).

It is only the first reaction, leading to the species absorbing at 590 nm, that is light-driven, the remaining transitions back to the native species (bR_{570}) being driven by thermal energy. There is evidence from circular and linear dichroism studies that the orientation of the retinal chain within the membrane shifts during the cycle, and that the conformation of the protein also undergoes a cyclic change. The Schiff's base appears

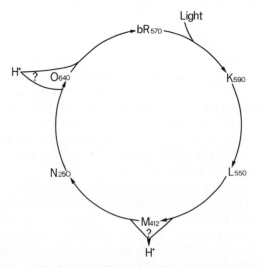

Figure 9.8. The photocycle of bacteriorhodopsin: bR_{570} is the form of bacteriorhodopsin found in purple membrane exposed to light and contains all-*trans* retinal. Flash illumination results in the rapid (psec) production of an intermediate K_{590} that thermally isomerizes to L_{550} and then to M_{412}. Proton release occurs at or immediately following the M_{412} stage and reuptake of a proton occurs at some point between N_{250} and bR_{570}. The thermal isomerizations leading from K_{590} back to bR_{570} occur in the millisecond time range.

to be directly involved in the proton uptake/release mechanism, and the retinal-lysine linkage is certainly protonated in bR_{570} and in K_{590} and unprotonated in M_{412}. There are some preliminary indications that a photoisomerization of bR_{570} in the region of the 14–15 bond in the retinal chain could transport the nitrogen-bound proton from the cytoplasmic face of the purple membrane to a part of the membrane in contact with the extracellular environment. It is suggested that, in the protonated form, re-isomerization is hindered but that release of the proton permits thermal re-isomerization to the all-*trans* state. The all-*trans* M_{412} species may then accept a proton from the cytoplasmic face

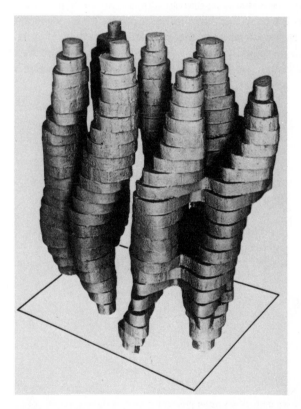

Figure 9.9. A model of a single bacteriorhodopsin molecule: the top of the model corresponds to the cytoplasmic side of the membrane: from R. Henderson and P. N. T. Unwin, *Nature* (1975), **257**, 28–32; original photograph by courtesy of Dr. Richard Henderson, M.R.C. Laboratory of Molecular Biology, Cambridge.

to re-form bR_{570}. Although the detailed steps of proton translocation are still the subject of considerable debate, there is universal agreement over the basic function of bacteriorhodopsin, which is to utilize light energy directly to pump protons from the cytoplasm to the extracellular environment of *H. halobium*.

Unwin and Henderson were able to construct a model of the arrangement of bacteriorhodopsin in the purple membrane using data from electron microscopy and electron diffraction studies. It is now known that this model, shown in figure 9.9, is an accurate representation of a single bacteriorhodopsin molecule, and that the top of the model corresponds to the cytoplasmic face of the membrane.

Thus we see seven rod-like structures, about 4 nm long and 1 nm apart, that traverse the membrane approximately perpendicular to its plane. It is presumed that the rods are the α helical portions of the protein. Unfortunately we still do not know the position of the retinal moiety, although we can look forward to this knowledge in the near future.

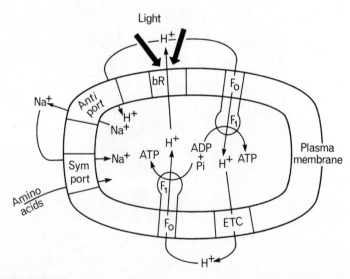

Figure 9.10. Membrane activities of *H. halobium*: Bacteriorhodopsin, bR, acts as a light-powered proton pump. The proton gradient is used to drive a number of processes: Na^+ is pumped out of the cell by an $H^+:Na^+$ antiport, the Na^+ gradient drives amino acid uptake via an Na^+ symport. ATP synthesis is also powered by the proton gradient. Under conditions where respiration can proceed the electron transport chain, ETC, generates the trans-membrane proton gradient.

Under the environmental conditions described at the outset, the $\Delta\tilde{\mu}_{H^+}$ generated by rhodopsin is the only energy source available to *H. halobium*. In this state, the purple membrane constitutes about 50% of the total plasma membrane area. The remainder of the membrane, referred to as *red membrane*, contains many different proteins and carries out all the functions normally associated with a plasma membrane. In the red membrane there is a coupling ATPase which is driven by $\Delta\tilde{\mu}_{H^+}$ to synthesize ATP. *H. halobium* also utilizes $\Delta\tilde{\mu}_{H^+}$ as the energy source for other processes (e.g. metabolite transport) as we saw in the previous chapter (page 186). Under aerobic conditions, the purple membrane is lost and $\Delta\tilde{\mu}_{H^+}$ is generated by the respiratory chain found in the red membrane. Figure 9.10 summarizes the overall activity of *H. halobium*.

The purple membrane is uniquely simple in its operation, and we might ask why the highly complex proton pumping systems that we discussed in earlier sections nevertheless prevail. The answer may be that the purple membrane is very inefficient, in that only 5–10% of the absorbed light energy is converted into electrochemical energy (as an H^+ gradient). Thus the successful operation of the membrane depends on very high light levels, which places a great restraint on the organism.

As stated previously, bacteriorhodopsin has a common chromophore with rhodopsin, the visual pigment, and there are some functional aspects that the two systems have in common. In the next section we shall consider the disc membrane of the retinal cell and examine the role of rhodopsin in the transduction of light energy into an electrical signal.

Rhodopsin and photoreception

The arrangement of the membraneous discs which are found in the outer segments of retinal rods was described in chapter 3. The overall function of the rod cells is to convert light energy into an electrical signal. The electrical signal, as in all biological systems, takes the form of a changed distribution of ions, and it is the task of rhodopsin to mediate the transduction of light energy into electrochemical energy.

Rhodopsin is a glycoprotein of molecular weight about 35 000, comprising a chromophore, 11 *cis* retinal, bound to the apoprotein, *opsin* by a Schiff's base linkage (with the ε-amino group of a lysine residue). The isolated protein has a high content of hydrophobic amino acids and can only be maintained in solution in the presence of detergents. Rhodopsin, as we might expect, is an integral protein of the disc membrane, and there are reports that it constitutes as much as 90% of the membrane protein.

There have been conflicting reports about its exact disposition. Early X-ray and neutron diffraction studies suggested that rhodopsin was located at the intra-discal face, but it is now generally agreed that rhodopsin spans the disc membrane; part of the peptide chain is exposed at the cytoplasmic face, and a glycopeptide tail extends into the intra-discal space. The intradiscal surface is derived from the outer face of the plasma membrane (chapter 3, page 36) and so the orientation of rhodopsin corresponds to that of other integral membrane glycoproteins (see figure 5.41 and chapter 7).

Rhodopsin is not found in the rigid crystalline arrays seen in the purple membrane (page 207). The disc membrane has an exceptionally high content of polyunsaturated fatty acids, which ensures a highly fluid environment. R. Cone and his colleagues have shown that the vertebrate disc membrane has a viscosity at least an order of magnitude lower than that of erythrocyte membranes. In this fluid medium, rhodopsin rotates about an axis perpendicular to the plane of the membrane and also undergoes rapid lateral diffusion. Linear dichroism studies reveal that the retinal projects laterally from the apoprotein in a plane almost parallel to that of the membrane, so allowing the spinning chromophore molecule to sweep out a large area of membrane. This arrangement clearly optimizes the absorption of light passing axially through the stacked discs.

In the dark there is a constant release of neurotransmitter at the synapse formed at the base of the retinal rod cell (chapter 3, figure 3.10a) and a train of impulses is generated in the optic nerve. Illumination of the disc membranes results in a transient reduction in the Na^+ permeability of the rod cell plasma membrane, which leads to a reduction in the release of neurotransmitter at the synaptic region, and a modulation of the steady signal of the dark state. How is communication between these two separate membranes effected? The currently accepted answer is that Ca^{++} ions provide the coupling factor. In the resting (or dark) state, Ca^{++} is retained within the discs. Illumination promotes the opening of a Ca^{++}-specific channel in the disc membrane, and Ca^{++} ions diffuse out into the outer-segment. The Ca^{++} ions are known to inhibit Na^+ entry through the outer segment plasma membrane and, because the Na^+, K^+-ATPase continues to operate, the rod cell becomes hyperpolarized (page 215) and transmitter release is diminished. The Ca^{++} flux would seem to function as an amplification system, as a very few Ca^{++} ions can inhibit the entry of many Na^+ ions. The question is, of course, whether rhodopsin is the light-sensitive Ca^{++} channel. There is good experimental evidence that rhodopsin alone can perform such a function, and defined phospholipid bilayer vesicles containing rhodopsin as the only protein do show a light-dependent Ca^{++} permeability.

The mechanisms by which rhodopsin may act as a variable Ca^{++} channel in the disc membrane are not clear. As in the case of bacterio-rhodopsin, rhodopsin also undergoes a series of light-induced conformational changes. The sequence of events in rhodopsin is similar, but nevertheless distinct from that of its bacterial counterpart. Illumination promotes an isomerization of the ground-state 11-*cis*-retinal to all-*trans* retinal which then dissociates from the apoprotein. There are known to exist four or five spectrally-distinct intermediates between the initial absorption of a photon and the release of all-*trans* retinal. Flash spectroscopy has yielded detailed information on the spectral characteristics of the intermediates but has provided little insight into the precise mechanisms of transduction. In the early stages of the photolysis sequence, changes in apoprotein conformation are seen with a time scale of milliseconds. Accompanying this change there is also a perturbance of a phospholipid molecule, probably phosphatidylethanolamine, which is known to be tightly bound to rhodopsin in the membrane. We can only speculate that the several light-induced changes permit the modified rhodopsin to function as a Ca^{++} channel. Indeed, it has been suggested that the apoprotein, freed from its chromophore side-arm, may so be enabled to undergo rotational movements in the membrane and to function as a diffusional carrier for Ca^{++}.

Following illumination, it is necessary to restore the disc membrane to its ground state, and we have some information about the various steps in this process. During dark adaptation, the retinal, released as all-*trans* retinal in the photolysis sequence is isomerized back to 11-*cis* retinal, but it is not clear exactly how or where in the rod cell this occurs. It is known that once 11-*cis* retinal is formed, rhodopsin is regenerated spontaneously. The bleached rhodopsin or opsin may be phosphorylated by an ATP- and/or GTP-requiring kinase, and this phosphorylation is light-dependent insofar as only bleached opsin or rhodopsin can act as substrate. The phosphorylated rhodopsin is less sensitive to light, and it may be that phosphorylation is the mechanism by which adaptation to high light levels occurs. There is indeed some evidence to suggest that the release of Ca^{++} from the discs may be affected by the degree of phosphorylation of the rhodopsin.

Accompanying the post-illumination recovery of rhodopsin is an active re-uptake of Ca^{++} by the discs. The presence of a Ca^{++}-dependent

ATPase in rod outer segments has been demonstrated, but the experiments did not establish that the enzyme is associated with the discs. That the enzyme is localized within the disc membrane is strongly suggested by experiments in which dark-adapted discs are seen to accumulate Ca^{++} in an ATP-dependent manner.

The retinal rods are the best understood of the several sensory receptors of the nervous systems. Other systems, such as touch, taste, smell, or sound receptors, must also depend on an initial transduction step, at which point the sensory information is converted to an electrical signal. As yet we have little information about the molecular mechanisms of such transducers. Within the nervous system, information is transmitted as electrical signals and in chapter 3 we discussed the specialized regions of the nerve cell, the synapses, at which intercellular communication occurs. The mechanisms of synaptic transmission are well understood, and we shall consider these in the next section.

Synaptic transmission

A fundamental prerequisite to the chemical theory of synaptic transmission is the existence of a specialized *receptor* at the surface of the postsynaptic membrane which recognizes the neurotransmitter (chapter 3, page 28). Transient interaction of the transmitter with the receptor initiates a change in the permeability characteristics of the membrane; for it is known that it is in such a permeability change that the 'message' carried by the transmitter is first expressed. In functional terms, therefore, we can consider that two separate events occur at the postsynaptic membrane; firstly, the transmitter interacts in a highly specific manner with its receptor and, secondly, a permeability change in the membrane occurs. The permeability change results in a redistribution of ions across the membrane, which promotes a marked change in $\Delta\psi$. Thus interaction of the transmitter with its receptor generates an electrical signal in the postsynaptic cell. At many synapses this change in electrical status of the postsynaptic membrane takes the form of a rapid inward current of sodium ions, followed by a slower outward potassium current, and is termed an *action potential*. Its characteristics, however, were first established not at synapses but in the axon.

The action potential

The ionic changes that give rise to the action potential were elucidated by a series of now classical experiments in the late 1940s and early 1950s by L. A. Hodgkin, A. F. Huxley and B. Katz. Most of these experiments were carried out using the giant axon of the squid. This preparation provides a single nerve fibre more than 0·5 mm diameter into which recording and stimulating microelectrode assemblies can be inserted. The membrane potential of the resting axon is about -65 mV (inside negative) and it was seen that when axons are depolarized (by injection of positively charged ions) the membrane potential initially passively follows the imposed depolarization. However, as $\Delta\psi$ approaches the *threshold value*, which is about -15 millivolts, a major change in the permeability properties of the axon plasma membrane occurs. There is a sudden increased permeability to Na^+ ions, leading to their rapid influx into the cell, so causing further depolarization. The influx of Na^+ continues until the membrane potential almost reaches E_{Na^+} (the equilibrium potential for Na^+, see page 179). This phase of the membrane response is a self-perpetuating process, because the continuing depolarization of the membrane further lowers the resistance to sodium flow. As E_{Na^+} is approached, however, a further series of changes occurs. Resistance to Na^+ entry increases, and is accompanied by a sharply decreased resistance to K^+ exit. The outward flux of K^+ halts the depolarization of the axon, and *hyperpolarization* of the membrane commences. The exit of K^+ continues until the original resting $\Delta\psi$ is approached. Invariably some overshoot occurs, and $\Delta\psi$ transiently increases beyond the normal resting level. The whole sequence constitutes the *action potential* and is summarized diagrammatically in figure 9.11.

The action potential is the form in which information is passed along the axons. The impulse travels as a wave of depolarization, each minutely contiguous section of the membrane being depolarized to its threshold value by the electrical events occurring adjacent to it, and then going through the sequence of events outlined in figure 9.11. The passage of action potentials along an axon leaves the neurone with a disturbed ion distribution, and it is at this point that the energy-requiring Na^+, K^+-ATPase comes into play and restores the ion balance. It is important to realize, however, that the action potential occurs as a result of the *passive movement* of ions across the membrane, i.e. the energy of the action potential derives from the concentration gradients of Na^+ and K^+ that are maintained by the Na^+, K^+-ATPase.

In the foregoing discussion we considered the initiation of the action potential as being a provoked depolarization of the axon membrane. The initial depolarization of the membrane is not achieved *in vivo* through the intervention of microelectrodes. We have briefly introduced the

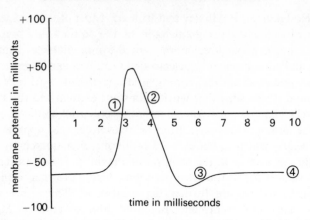

Figure 9.11a. The action potential: The main features are: (1) a rapid rising phase; (2) a slower falling phase, combining to give a *spike* of duration about 1 millisecond; (3) a longer hyperpolarization lasting several milliseconds; (4) recovery of the resting potential.

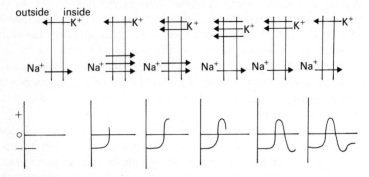

Figure 9.11b. Trans-membrane movements of K^+ and Na^+ with the development of the action potential.

concept of a neurotransmitter-induced change in membrane permeability, and the precise mechanisms by which this process occurs have occupied neurobiologists since the early 1900s. Much of the work has concentrated on the vertebrate neuromuscular junction.

The neuromuscular junction

This system offered two advantages to early investigators. Firstly, the chemical nature of the neurotransmitter, acetylcholine, was established as early as 1920 and, secondly, the overall transmitter-induced response of the postsynaptic membrane is easily measured, as it takes the form of a contraction of the muscle. In the ensuing years, electrophysiological experiments have permitted the construction of a detailed picture of the events following acetylcholine-receptor interaction.

The initial interaction of acetylcholine with its receptor produces small localized changes in $\Delta\psi$, known as miniature end plate potentials or m.e.p.ps. These arise as the result of the opening of discrete sodium channels that are closely associated with the receptor molecules and that allow Na^+ ions to flow across the postsynaptic membrane.

In the presence of larger quantities of acetylcholine, e.g. the quanta described in chapter 3, many m.e.p.ps. are generated simultaneously and summate, thereby significantly shifting $\Delta\psi$ of the membrane at and around the synapse. As $\Delta\psi$ approaches the threshold value, so an action potential is generated and spreads over the entire muscle plasma membrane, and a contraction follows (see chapter 8, page 183).

At the nerve-muscle synapse, therefore, the generation of the action potential in the postsynaptic cell may be considered as two separate events. Firstly, transmitter-receptor interactions lead to limited Na^+ flux and the generation of m.e.p.ps.; secondly, Na^+ channels in the plasma membrane open and an action potential is produced.

We are fast approaching the stage where we have a good understanding of the molecular events occurring at these two stages.

The acetylcholine receptor protein

The acetylcholine receptor at the vertebrate neuromuscular junction is classified pharmacologically as a *nicotinic* cholinergic receptor. This means that the receptor not only responds to the natural transmitter but also to the compound *nicotine*, and in this way is distinguishable from other cholinergic receptors that respond to *muscarine*. Such *muscarinic* cholinergic receptors are found primarily in the brain and on smooth muscle cells. The two cholinergic systems differ in the speed at which they react—the nicotinic type being characterized by a rapid 1–3 millisecond response compared with times of over 100 milliseconds required by muscarinic synapses. It seems that the molecular events following acetylcholine-receptor interaction are quite different in the two systems, and for the moment we shall confine our discussion to the nicotinic receptor.

Until recently the very low levels of acetylcholine receptor present in

10 nm

Figure 9.12a. Electron micrograph of a freeze-etched preparation of receptor-rich membranes from electroplax of *Torpedo marmorata*. A hexagonal array of 'rosettes' is seen, and the substructure of the individual particles can be seen in some cases (arrows). The appearance of the rosettes should be compared with the negatively stained preparation of purified receptor protein from the electroplax of *Electrophorus electricus* (figure 9.12*b*). The rosettes are seen to consist of 5 or 6 sub-units some 3 to 4 nm in diameter, with a central pit into

the plasma membrane of vertebrate muscle cells (less than 1% of the total plasma membrane protein) precluded its isolation by conventional biochemical techniques. Instead, therefore neurochemists turned their attentions to an unusual tissue, the electric organ of various species of electric fish. The electric organs are made up from thousands of identical cells or *electroplaques* that are embryologically derived from muscle cells. During development, the cells lose their contractile elements, but undergo a great proliferation of the synaptic region. Thus the adult electroplaque

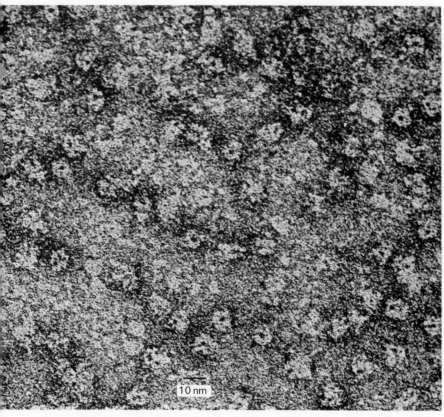

which the stain has penetrated. Figure 9.12*a* is from J. Cartaud, E. L. Benedetti, J. B. Cohen, J.-C. Meunier and J.-P. Changeux, *FEBS Lett.* (1973), **33**, 109–113. **Figure 9.12b** is from J.-C. Meunier, R. Sealock, R. Olsen and J.-P. Changeux, *Eur. J. Biochem.* (1974), **45**, 371–394. Both original micrographs by courtesy of Dr. Jean Cartaud, Institut de Biologie Moleculaire, Paris.

is a large multinucleate cell, one face of which is virtually entirely post-synaptic membrane. The receptor density in this membrane is high (approaching 25% of the total membrane protein) and it constitutes an excellent source for isolation of the receptor, which appears to be pharmacologically identical to that of the vertebrate muscle cell.

A further invaluable aid to the isolation of the receptor is the fact that several snakes of the cobra and krait families produce in their venom polypeptides that interact in a highly specific manner with the nicotinic receptor. The peptides, known as α-toxins, appear to bind almost irreversibly to the acetylcholine binding site of the receptor with dissociation constants (K_d) of the order of 10^{-12}M. Physiologically the effect of injection by such toxins is a chronic and severe paralysis of the skeletal musculature. The α-toxins provide two powerful tools in receptor isolation studies. Firstly, the radiolabelled toxins may be used to detect the presence of receptors in various preparations of membrane proteins and, secondly, they can provide specific ligands for affinity chromatographic (chapter 5) isolation methods.

The acetylcholine receptor protein has been isolated in a highly purified form from electroplaques. It is an integral membrane protein of molecular weight 250 000–300 000, and appears in the electron microscope as a rosette-like structure comprising five or six sub-units (see figure 9.12*b*). Freeze-fracture of the innervated face of the electroplaques shows these same structures to be present (figure 9.12*a*).

There is still some disagreement among the various research groups about the numbers and sizes of the sub-units, although the differences may to some extent be species-related. There is, however, a general concurrence that in all cases the acetylcholine binding site is confined to a single class of peptide having a molecular weight of about 40 000.

More recently, the skeletal muscle receptor has been isolated and found to be similar to the fish protein. Apart from the basic academic interest in the acetylcholine receptor protein, the last five years have also produced a major research activity concerned with a medical aspect of the receptor. This stems from the realization in the mid-1970s that myasthenia gravis, a muscle disease characterized by weakening and rapid tiring of the skeletal musculature, is an *autoimmune* disorder in which the nicotinic acetylcholine receptor is the major autoantigen. The myasthenic patient produces antibodies that are directed against the receptor protein and that bind to the receptors at the muscle synapse. This may have a dual effect: firstly, the receptors are effectively reduced in numbers and thus transmission is impaired; secondly, the bound antibody may initiate a cell-mediated attack on the muscle surface membrane, leading to its destruction. It is important that we have a detailed knowledge of the immunochemical characteristics of the receptor protein in order that its role in the pathogenesis of myasthenia gravis may be understood. Many laboratories are working towards this goal and it seems probable that in the next few years the protein will be completely characterized.

What of the functional aspects of this protein? Can we formulate a model of receptor-mediated Na⁺ flux? A clear picture has not yet emerged, and there are equivocal findings reported in the literature. The basic question is whether the Na⁺ channel or ionophore is an integral part of the receptor protein, or whether it is a separate molecular entity. Figure 9.13 summarizes the alternatives.

Early studies suggested that the receptor protein prepared by affinity chromatography of detergent-solubilized electroplaques membranes

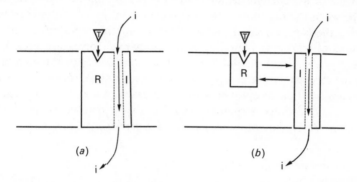

Figure 9.13. Receptor-ionophore coupling: in scheme (*a*) the receptor R and Ionophore I form a single membrane unit. Interaction of transmitter T with R results in a conformational change such that I opens and allows ions i to flow through. In (*b*) R is a discrete membrane constituent separate from I. The information carried by T and expressed as a conformational change in R must be transmitted through the membrane to I.

carries within it the ionophore component. Lipid vesicles into which such receptor preparations were incorporated showed an acetylcholine-dependent Na⁺ flux. Other studies showed that a receptor preparation prepared by organic solvent extraction of electroplaques (i.e. the receptor was prepared as a proteolipid, chapter 5) conferred acetylcholine-dependent sodium permeability to black lipid films.

The recent availability of a specific natural label for the ionophore has prompted a further examination of the problem. The compound *histrionicotoxin* is a steroid-like molecule produced by certain species of Colombian tree frogs. Its physiological action is to prevent the formation of m.e.p.ps. at the vertebrate neuromuscular junction. The toxin does not affect electrically-provoked action potentials, and it can be shown that it does not compete with the snake α-toxins for the acetylcholine binding site. By using radiolabelled histrionicotoxin, various research groups have sought to identify the ionophore component of the electroplaques membrane. The currently available data lead to conflicting conclusions. On the one hand there are reports that highly purified receptor protein does not carry a histrionicotoxin binding site, but that a separate membrane peptide of molecular weight 43 000 has such a site. In contrast is the finding that receptor preparations initially contain a 43 000 chain that can be released by alkali treatment, but that the histrionicotoxin binding of the receptor is not affected by this treatment.

In spite of the current debate about the precise nature of the relationship between the ionophore and the receptor in the case of the isolated protein, there is universal agreement that in the membrane the two functions of transmitter binding and ion channel formation are closely coupled. Freeze-fracture and X-ray diffraction studies of the membrane show the receptors to be closely packed in a crystal-like hexagonal array. It seems likely that receptor-ionophore interaction may be dependent on such a highly ordered structure, and that this is completely lost in the case of the detergent-solubilized isolated receptor.

The ion gates of the action potential

The ion gates or channels that are associated with the action potential are quite distinct from the ionophore of the receptor. As we have seen in an earlier section (page 215), the action potential can be divided into two separate phases. Firstly, a very rapid influx of sodium occurs, followed by a rather slower efflux of potassium. The permeability of the channels is regulated by $\Delta\psi$, changes in which can be promoted by the receptor-mediated generation of m.e.p.ps. or, under experimental conditions, by way of microelectrodes.

The existence of voltage-regulated gates in the axon membrane was first postulated by Hodgkin and Huxley in the early 1950s, and their early electrophysiological experiments, some of which were carried out in conjunction with Katz, confirmed their suggestion that discrete ion gates are present in the membrane and that the gates discriminate between sodium and potassium ions.

The technique devised by Hodgkin and Huxley to make their quantitative measurements on the ion channels is known as *voltage-clamping* and is perhaps the most powerful tool that we have for studying excitable membranes. Basically the technique involves holding or clamping $\Delta\psi$ at a fixed value. This is achieved by inserting two electrodes into the cell, one of which records the internal potential, whereas the other passes current across the membrane to a third extracellular electrode. The membrane potential is monitored and compared constantly with a control voltage set by the experimenter. A feedback amplifier allows sufficient current to flow across the axon membrane to equalize the two values. Thus the membrane potential can be fixed at a chosen level, and the voltage dependence of any transmembrane ionic currents can be measured. The voltage can also be moved virtually instantaneously to a different level, and the ensuing changes in membrane current measured.

The selectivity of the ion gates in voltage-clamped axons was examined by Hille. He found that the size of the ions determined whether they could enter the channels through which they proceeded in single file. Hille's experiments suggested the presence of an energy barrier at the entrance to the channel, to which the name *selectivity filter* has been given. In the case of the Na^+ ion, passage through the selectivity filter involves loss of the hydration shell, but the larger K^+ ion cannot achieve this loss of water so easily, and thus encounters a much higher energy barrier. This situation is reminiscent of that seen in the cation-selective antibiotic ionophores, where the ionophore provides a critically-sized replacement environment for the hydration shell (chapter 8, page 171).

Natural toxins have been of great help in elucidating the mode of operation of the ion gates. *Tetrodotoxin* is a compound found in the tissues of puffer fish, the California newt and the Australian blue-ringed octopus. A very similar molecule *saxitoxin* is found in a marine plankton *Gonyaulax*. Both toxins prevent the formation of action potentials by blocking the Na^+ channels. It has been demonstrated that the toxins bind to the outside of the channel, as high concentrations of toxin injected into axons have no effects on action potentials. The stoichiometry of binding is one molecule of toxin per channel, and radiolabelled toxin has been used to count the number of Na^+ channels per unit area of axon membrane. The values range from about $13/\mu m^2$ to $500/\mu m^2$, depending on the species, and correlate well with the values predicted from Na^+ flux measurements.

Hodgkin and Huxley reasoned that the dependence of the ion gates on the membrane potential means that the gates themselves are charged membrane components that move from a closed to an open state as the membrane potential reaches the threshold voltage. The movement of such charged entities within the membrane would generate a current, and Hodgkin and Huxley predicted the existence of such *gating currents*. However, they were never able to measure them, as they were completely masked by the much greater ionic currents.

It has recently proved possible to measure the gating currents directly, and Eduardo Rojas in collaboration with Richard Keynes has made many such determinations. The detection of the gating currents depends on the complete elimination of the much larger ionic currents, and it is in this respect that tetrodotoxin has been particularly useful. The Na^+ current can be blocked absolutely by applying tetrodotoxin, which fortunately does not appear to hinder the movement of the gate components. Potassium channels are blocked by replacing the internal potassium with caesium or tetramethylammonium ions. In such membrane preparations, it is not possible for transmembrane ionic currents to flow but, when the membranes are voltage-clamped and the potential is suddenly changed, a small current is detected. The current rises to a peak immediately after the change in potential, and then falls exponentially as the gating particles come to rest; computer-averaging of repeated responses has allowed the construction of an acceptable model for channel formation. Rojas proposes that the sodium channel is not a permanent structure, but that aggregates of integral proteins are present in the membrane which come together under the influence of $\Delta\psi$ to form a functional channel.

The selectivity filter is considered to be a protein that protrudes from the membrane surface and is the component to which tetrodotoxin binds. There is some indirect evidence that this component has a molecular weight of 200 000. The remainder of the channel is thought to be made up from integral proteins of molecular weight about 100 000, each having a large dipole moment. The functional channel has a molecular weight of about 500 000.

So far it has not proved possible to make direct measurements of the gating currents associated with the formation of the potassium channels. This is due in part to the lack of such a potent blocking agent as tetrodotoxin, and also to the relative slowness of opening of the channels. It is presumed, however, that an analogous voltage-dependent system operates, and we can anticipate the elucidation of its characteristics in the near future.

Electrophysiological measurements have thus provided a particularly clear picture of the functioning of the ion gates that give rise to the action potential. Biochemical studies of the channels are not so far advanced. The electric organ of the electric eel *Electrophorus electricus* has a particularly high capacity to generate action potentials, and work is in progress to isolate the channel proteins. A tetrodotoxin-binding protein of molecular weight about 200 000 has been isolated from this source. The hope is that the isolated proteins can be incorporated into lipid bilayers, and that voltage-dependent ion permeability can be studied in defined model systems.

Although the measurements of gating currents have all been made on axon membranes, it seems probable that the same mechanisms provide for the generation of action potentials at the nicotinic neuromuscular junction, and we can speculatively formulate a model that provides a complete account of an acetylcholine-initiated action potential (figure 9.14).

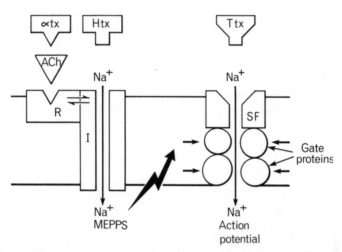

Figure 9.14. The initiation of an action potential at a nicotinic cholinergic synapse: Acetylcholine ACh interacts with the receptor R; the interaction results in an opening of the ionophore I, and Na^+ ions enter the cell generating MEPPS. The summated MEPPS change $\Delta\psi$ and promote an aggregation of the intrinsic gate proteins with the selectivity filter SF, thereby forming a functional sodium channel. The system can be blocked at three distinct sites by toxins: αtoxin, αtx; histrionicotoxin, Htx; tetrodotoxin, Ttx.

Other neurotransmitter receptors

The nicotinic cholinergic receptor is the best characterized of the neurotransmitter receptor systems. The receptors for other neurotransmitters have not yet been isolated in a purified state, although there is currently a great deal of interest in assaying the membrane-bound receptor sites by measuring the specific binding of radiolabelled ligands. Progress in the isolation of the receptors would be greatly aided by highly specific tight-binding toxins analogous to the nicotinic α-toxins. It seems also that the nicotinic system may be exceptionally simple with respect to its molecular structure, in that transmitter binding and the subsequent changes in membrane permeability are probably achieved by a single polymeric protein. Thus, for example, the muscarinic receptor shows little resemblance to its nicotinic counterpart. It appears to comprise a readily extractable membrane protein that is thought to carry the acetylcholine binding site, but there has been little success in isolating or characterizing this protein. The receptor does not regulate permeability directly, but increases membrane permeability to Ca^{++}. The influx of calcium ions is thought to affect intracellular cyclic nucleotide levels, and the cyclic nucleotides in turn may modulate channel-forming membrane proteins.

Receptors and cyclic nucleotides

The role of cyclic nucleotides as modulators of intracellular metabolism was elucidated from studies of hormone systems, and we shall consider

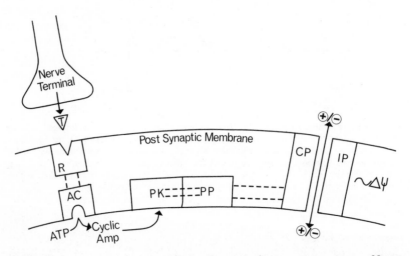

Figure 9.15. Involvement of cyclic AMP in neurotransmitter-receptor responses: Neurotransmitter T is released from the nerve terminal and interacts with the receptor R in the postsynaptic membrane; the transmitter-receptor complex activates the membrane bound enzyme, adenylate cyclase AC, which produces cyclic AMP at the cytoplasmic face of the membrane. Cylic AMP modulates the activities of membrane-associated protein kinase-phosphoprotein phosphatase systems PK, PP that control the activity of channel proteins, CP, or electrogenic ion pumping proteins IP; thus Δψ changes and the postsynaptic cell may be either depolarized or hyperpolarized. After P. Greengard, *Nature* (1976), **260,** 101–107.

this work in a later section (page 229). They are also implicated in some types of synaptic transmission and their involvement in mediating the effects of catecholamine neurotransmitters has been studied in particular detail. A generally-accepted sequence of the events following transmitter-receptor interaction in this system has emerged and is shown in figure 9.15.

A further aspect of the neurotransmitters that we have not considered is that responses may be either *excitatory* or *inhibitory*. Excitatory transmitters result in a depolarization of the membrane and the generation of an action potential (e.g. the nicotinic cholinergic system). Inhibitory transmitters produce a hyperpolarization of the membrane and thus render the production of an action potential by the postsynaptic cell less likely. In many cases the inhibitory systems are seen to involve a chloride channel that allows extracellular chloride ions to enter the postsynaptic cell. It is not uncommon to find a transmitter having either excitatory or inhibitory effects depending on the cell type. In such cases the receptor systems must differ. In the case of the cyclic nucleotide-mediated systems, distinct excitatory and inhibitory receptor proteins may interact with a common enzymic unit. Thus the catecholamine transmitter, dopamine, activates the membrane-bound adenylate cyclase, whereas the receptors for the opiate peptide transmitters are also coupled to adenylate cyclase, but in this case occupancy of the receptor leads to an inhibition of the enzyme.

The excitable membrane of *Paramecium*

In earlier sections we have seen how useful simple organisms, such as bacteria, can be as natural models in which to study complex membrane phenomena. The excitable membranes of the nervous system present a particularly daunting challenge to the membrane researcher, and the availability of a simple model is an attractive prospect. Such a system may be provided by the unicellular animal *Paramecium*.

Paramecium swims by beating its cilia, and the direction of the ciliary beat is controlled by the electrically-excitable plasma membrane. When normal paramecia come into contact with an obstacle or an undesirable solute, they avoid it by reversing the direction of the ciliary beat, thereby swimming backwards for a short distance, and then swimming forwards again in a different direction. It has been demonstrated that a small depolarization of the plasma membrane initiated by physical contact or changed ionic environment results in a graded action potential of similar form to that seen in axon and muscle membranes. However, the initial inward current in paramecia is carried by Ca^{++} and is followed by a slower outward K^+ current. The increased internal Ca^{++} concentration causes the cilia to reverse their direction of beat by an as yet unknown mechanism.

The Ca^{++} current channels are located exclusively in the portions of the plasma membrane covering the cilia, as removal of cilia eliminates the inward Ca^{++} current; this returns over a period of several hours as the cilia regrow and the plasma membrane is regenerated. *Paramecium* offers the same attractions as do other micro-organisms. It can be mass cultured, cloned and manipulated genetically. This last property has been exploited, particularly with reference to the Ca^{++} channel. A mutant type known as a *pawn* mutant (named after the chess piece—they can only swim forwards) has been studied electrophysiologically and found to have functionally defective Ca^{++} channels. Voltage-clamp experiments show that the *pawn* mutant can produce only the outward K^+ current, and comparisons of the current characteristics of normal membranes and *pawn* membranes have greatly facilitated the resolution of the overall action potential into its constituent inward and outward currents. It seems likely that, as in the case of transport studies (chapter 8), the use of micro-organisms such as *Paramecium* may allow us to elucidate the fundamental processes of membrane excitability and to extrapolate these findings to the inordinately complex nervous systems of the higher animals.

Hormone receptors

The neurotransmitters are a diverse group of molecules that act through various mechanisms to achieve a common goal, namely, to modulate the ion permeability of the postsynaptic membrane. Hormones are a similarly diverse group of compounds that also interact in many cases with cell-surface receptors. The physiological effects of these interactions, however, impinge upon a far wider range of cellular activities than is the case with the neurotransmitters. We have considerable knowledge of hormone-receptor interaction, and in some cases the initial stages of receptor-response coupling are understood. For many hormones, both of these activities are associated with the plasma membrane, and cyclic nucleotides are commonly involved.

Before proceeding to discuss these systems, it should perhaps be pointed out that one major group of hormones, the steroids, do not exert their physiological effects via a plasma-membrane receptor system. The steroids are probably unique in entering their target cell (passage across the plasma membrane is facilitated by their hydrophobic nature). Once inside the cell, the hormones interact with a cytoplasmic receptor protein, and the hormone-receptor complex enters the nucleus, where it is thought to modulate gene expression.

Hormones and cyclic nucleotides

Our understanding of the involvement of cyclic nucleotides in hormone action stems from the work of Sutherland and Rall in the late 1950s; they investigated the activation of liver glycogen phosphorylase by the catecholamine hormones (noradrenalin and adrenalin) and the peptide hormone glucagon. They showed that the activation of the phosphorylase in liver homogenates is a two-step process. Firstly, the addition of the hormones to the particulate fraction of the homogenate produces a small-molecular-weight heat-stable factor and, secondly, addition of this factor to the supernatant fraction stimulates glycogen phosphorylase. The heat-stable component turned out to be cyclic AMP. Sutherland and his colleagues formulated a general mechanism for hormone action that introduced the concept of first and second messengers. The hormone was

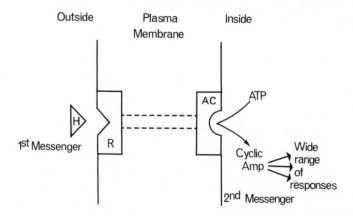

Figure 9.16. The second messenger hypothesis: the circulating hormone H is considered to be the 1st messenger, and conveys information to the outer surface of the plasma membrane where interaction with a specific receptor R occurs. The occupied receptor activates the membrane-bound enzyme adenylate cyclase AC, which produces cyclic AMP. This molecule constitutes the 2nd messenger and may affect a wide variety of intracellular activities, thereby giving rise to the overall response to the hormone.

seen as the first messenger, its job being to carry information to the plasma membrane of the target tissue, and there to interact with the receptor. The receptors are closely associated with a membrane-bound enzyme, adenylate cyclase, that produces cyclic AMP. Cyclic AMP acts as the second messenger and transfers the information to the internal machinery of the cell; the scheme is summarized in figure 9.16.

Sutherland and his colleagues, Robison and Butcher, suggested that cells might use second messengers other than cyclic AMP. There is increasing evidence that this is so, and cyclic GMP is thought to have such a role, as are calcium ions.

The mechanisms by which cyclic AMP affects the metabolism of cells are not completely understood. Activation of protein kinases is the only well-established intracellular effect, and it has been suggested that the whole diverse range of responses to cyclic AMP is mediated solely by these enzymes. Whereas it is probably true that this is the major mechanism, there is evidence that in many cells cyclic AMP-binding proteins other than protein kinases can affect the activity of enzymes.

The breakdown of cyclic AMP in the cell is achieved by the enzyme *cyclic AMP phosphodiesterase* which cleaves the internal cyclic diester to produce 5'-AMP. The metabolic activity of cells can be further modulated either by activating or by inhibiting the diesterase. However, it is unlikely that regulation of the enzyme is a major factor in the hormonal responses of cells.

Receptor-cyclase coupling

There is now general agreement that adenylate cyclase and the hormone receptors are distinct membrane proteins. Much of the information that we have derives from work on the β adrenergic receptor. The adrenergic receptors respond to adrenalin and noradrenalin, and are divided into two groups, α and β, on the basis of the tissue response or on pharmacological grounds. The α receptors are generally associated with involuntary smooth muscle contraction. Thus blood vessels are constricted, as are the spleen and uterus, but in contrast the smooth muscle of the intestine is relaxed by α receptor activity. Relaxation of involuntary smooth muscle is usually associated with β receptor activation. Thus blood vessels are dilated and bronchial smooth muscle is relaxed. However, the heart is stimulated by β receptor activation. Pharmacologically the receptors can be classified by their response to a range of adrenergic drugs. Phenylephrine acts at α receptors and isoprenaline is specific for

β receptor sites. The β effects are in general associated with activation of adenylate cyclase, whereas the α effects are not.

Erythrocytes from many species carry β receptors which have been studied in particular detail. An especially elegent series of experiments demonstrating the distinction between the β receptor and adenylate cyclase has been reported by Schramm and his colleagues. They inactivated the adenylate cyclase activity (by heat treatment or by N-ethylmaleimide) of turkey erythrocytes and fused them with mouse erythroleukemia Friend cells. These latter cells have no β receptor sites but do have adenylate cyclase activity. Within minutes of fusion, catecholamine-stimulated adenylate cyclase activity could be demonstrated, showing that the erythrocyte β receptors had established functional contact with the Friend cell adenylate cyclase. Other workers have shown that it is possible to separate β adrenergic binding activity from adenylate cyclase activity in solubilized frog erythrocyte membranes.

In addition to a membrane receptor protein and adenylate cyclase, the β adrenergic receptor response is dependent on GTP. The GTP is bound to a protein which is associated with the adenylate cyclase and which shows β-adrenergic-stimulated GTP-binding and GTPase activity. Fractionation studies on pigeon erythrocytes have shown that the GTP-

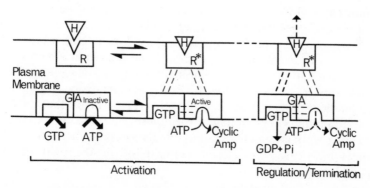

Figure 9.17. Hormone-mediated regulation of adenylate cyclase activity. In the left-hand part of the diagram activation of the cyclase is summarized: binding of hormone H to receptor R promotes interaction between the occupied receptor R* and the GTP binding protein/GTPase-adenylate cyclase complex GA. Coupling between R* and GA promotes GTP binding, which leads to activation of the cyclase. In the right-hand part of the figure, regulation/termination of the cyclase is shown; activation of the GTPase activity of GA results in the breakdown of bound GTP, which reduces cyclase activity; this may be initiated by the dissociation of H from R*.

binding protein can be separated from adenylate cyclase, and that under such conditions the adenylate cyclase no longer responds to GTP.

It is suggested that the hormone activation of adenylate cyclase is mediated through the GTP-binding proteins. The precise mechanisms by which such a regulation of the cyclase activity is achieved is not yet clear, but it has been suggested that adenylate cyclase is present in two forms having high and low activity respectively. Hormone binding may lead to formation of a GTP-protein complex which would stabilize the high activity form and increase cyclic AMP production. Regulation/termination of hormone action could then be brought about by activation of the GTPase activity of the binding protein, which would lead to release of GDP and a return of the cyclase to a low activity state. Support for this interpretation comes from the observation that guanyl 5'-yl imidodiphosphate (GppNHp), an analogue of GTP that cannot be hydrolysed, activates adenylate cyclase and potentiates the response to hormones in a variety of cells. Thus GppNHp binds to the GTP-binding protein but, as it is not a substrate for the GTPase activity, a chronic high activation of the cyclase ensues. Figure 9.17 summarizes current thoughts on the system.

Cholera toxin (see chapter 7, page 155) has greatly aided our understanding of the role of the guanine nucleotides in adenylate cyclase activation. Cholera toxin irreversibly activates adenylate cyclase in any mammalian cell having the toxin receptor and adenylate cyclase. The toxin molecule has two different sub-units, A and B; B is involved in binding to the ganglioside receptor (chapter 7, page 155), whereas A is concerned with activation of the cyclase. It has been shown in broken cell preparations that the enzyme can be activated by sub-unit A alone, which is now known to catalyse the ADP-ribosylation (from NAD) of the GTP-binding protein. Thus it is attractive to postulate that on ADP-ribosylation the GTP-binding protein loses its GTPase activity, so preventing termination of hormone-stimulated cyclase activation. These changes may cause the altered permeability of the tight junctions discussed on page 193.

Hormone receptors, Ca^{++} and cyclic GMP

An important regulatory component of the hormone-receptor-cyclase system is the intracellular Ca^{++} concentration. High levels of intracellular calcium inhibit the cyclase, and this may be the basis for the antagonism often seen between hormones that increase plasma-membrane Ca^{++} permeability (e.g. muscarinic cholinergic receptors, page 226) and those hormones such as catecholamines that activate adenylate cyclase. Calcium also inhibits the phosphodiesterase, an effect that is apparently antagonistic (with respect to the final concentrations of cyclic AMP) to the effects on the cyclase.

An endogenous calcium-binding protein *calmodulin* is thought to mediate these effects. Calmodulin is found in many tissues and is reported to have an involvement in the calcium-promoted phosphorylation of the contractile proteins of muscle, an activity that is distinct from its role in the cyclic nucleotide systems. It has recently been suggested that in brain calmodulin promotes an interaction between the catecholamine receptor and adenylate cyclase and/or the GTP binding protein.

In the early 1970s it was reported that acetylcholine raises cyclic GMP levels in heart muscle, and since then there has been much discussion about a possible second messenger role for cyclic GMP. Many groups have sought to demonstrate that the guanylate cyclase-cyclic GMP system is directly analogous and antagonistic to the cyclic AMP system, but there is still little experimental data to support this supposition. There is, however, no doubt that intracellular levels of cyclic GMP vary in response to hormonal activation, and that the changes are frequently opposite to those seen in cyclic AMP levels. There is no good evidence that cyclic GMP activates protein kinases, and it appears that in many cases observed hormonally-induced rises in cyclic GMP are the results of an increased intracellular Ca^{++} concentration. Calcium is known to activate guanylate cyclase and many hormones that lead to increased cyclic GMP levels increase plasma membrane permeability to Ca^{++} ions. Thus activation of the α adrenergic receptors in many tissues leads to a calcium influx, followed by an increase in cyclic GMP levels. It may well be that cyclic GMP is a second messenger, but at present we do not have a clear understanding of what the message is.

In many tissues activation of the receptors that regulate Ca^{++} permeability is accompanied by an enhanced turnover of phosphatidyl-inositol. Under these conditions the phospholipid is cleaved to yield inositol phosphate and a diglyceride. There is some evidence that the breakdown of phosphatidylinositol is linked to the hormone-receptor interaction, and that this is followed by increased Ca^{++} flux. It is not yet clear exactly how phosphatidylinositol breakdown is linked to Ca^{++} entry, and we can look forward to a clearer understanding of the interplay between this membrane phospholipid, calcium ions and the cyclic nucleotides.

Receptor mobility

We have already discussed the evidence supporting the suggestion that adenylate cyclase and the hormone receptors are distinct membrane proteins. There is also good evidence that in many tissues there is a considerable molar excess of receptors over adenylate cyclase. Thus, for example, the adenylate cyclase of the fat cell is stimulated by at least seven different hormones and is inhibited by three others. It is difficult to envisage ten discrete receptors, each with a recognition site at the

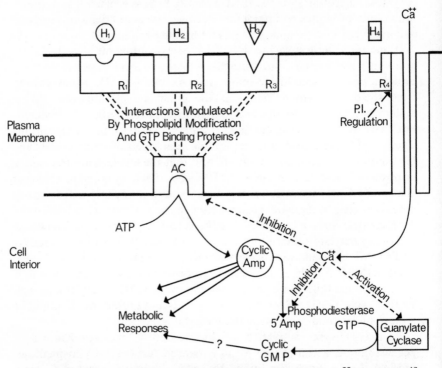

Figure 9.18. Cyclic nucleotide-mediated hormone receptor systems. Hormone specific receptors R_1, R_2, R_3 are free to migrate laterally in the membrane. Occupied receptors show enhanced interactions with adenylate cyclase AC; the interactions may be modulated by changes in the fluidity of the membrane (occasioned by modulation of the phospholipids) and by GTP binding proteins. Elsewhere in the membrane a receptor R_4 controls the calcium permeability of the membrane, possibly via the breakdown of phosphatidyl inositol PI. The entry of Ca^{++} into the cell may modulate the response to hormones H_1, H_2, H_3, as Ca^{++} is known to affect adenylate cyclase, phosphodiesterase and guanylate cyclase. Cyclic GMP may act antagonistically to cyclic AMP, but this is not well substantiated.

membrane surface, and each contiguous with the catalytic sub-unit of the cyclase. Pedro Cuatrecasas proposed that hormone receptors are not physically associated with the cyclase in the resting or non-activated state. He proposed a mobile receptor hypothesis in which hormone receptors diffuse laterally within the membrane. It is suggested that a receptor-hormone complex has a higher affinity for adenylate cyclase than the free receptor. Thus formation of a hormone-receptor-cyclase complex would occur at a rate dependent on the concentration of the receptor-hormone complex. Some receptor-hormone complexes would activate the enzyme, whereas others would inhibit it. In support of this interpretation of receptor-cyclase interaction it has been shown that temperature-dependent lag periods can be detected in the effects of some hormones on the cyclase.

There is evidence that coupling between the receptor and the cyclase is affected by the fluidity of the membrane and, in the case of the β adrenergic receptor in brain, it has been reported that binding of transmitter to the receptor promotes methylation of phosphatidyl-ethanolamine, yielding phosphatidyl-N-methylethanolamine. This results in increased membrane fluidity and enhances receptor-adenylate cyclase interactions. Thus we might speculate that membrane-associated methyl transferases may in part constitute the regulatory coupling factors that are thought to mediate the interaction between receptors and adenylate cyclase. Again these findings emphasize the regulatory role that membrane lipids may exercise over functional membrane proteins (see chapter 8, page 184). Figure 9.18 summarizes our current understanding of the cyclic nucleotide-mediated hormone receptor systems.

Receptor regulation

In the foregoing discussion we have considered the interaction between the hormone and its receptor to be of a transient nature, thus dissociation of the hormone leaves an unoccupied receptor site at the membrane surface that could interact with a further hormone molecule. There is, however, increasing evidence that this simple interpretation cannot satisfactorily explain the effects of some hormones on the target cell. Many hormone receptors seem to be regulated by the hormone or by other hormones. This is particularly well established for the peptide hormones. Thus, for example, increased insulin concentrations cause a decrease in the number of insulin receptors on liver cells, an effect termed *down-regulation* that has also been reported to occur for glucagon, growth hormone and the catecholamines. An important factor in the regulation of receptor numbers in this way is the incomplete dissociation of the hormone-receptor complex, and there is preliminary evidence that in some cases the binding of the hormone leads to a conformational change in the receptor, such that the binding affinity is increased. There is also evidence that, in the case of the cyclic AMP-mediated hormones, the association between the receptor and adenylate cyclase promotes the formation of the high-affinity state of the receptor.

The extended occupation of receptors by hormones has two effects: firstly, there is an immediate reduction in the number of free receptor sites, and therefore a reduction in target-cell sensitivity; secondly, prolonged occupation of the receptor is thought to lead to internalization of the complex by an endocytotic mechanism. There is now good evidence for the presence in cells of several hormones that are known to bind initially to receptors at the external surface of the plasma membrane. It is suggested that internalization is the major mechanism for the down-regulation of receptors that frequently follows exposure to high concentrations of hormone.

The biological role of the processes of receptor internalization and subsequent breakdown is not yet clear, apart from the obvious one of terminating the action of tightly bound hormones. The regulation of receptor numbers in this manner may also provide a means of protecting the cell from excessive stimulation and could be involved in the long-term hormonal regulation of cell growth and differentiation.

SUMMARY

1. Coupling or energy-transducing membranes generate proton gradients, the energy of which is utilized to synthesize ATP. The membranes contain a coupling ATPase that is forced to operate as an ATP synthetase by the prevailing proton gradient.

2. The primary energy source for the coupling membranes may be light, as in the thylakoid membranes of chloroplasts, or it may be oxidative metabolism, as in the inner mitochondrial membrane and in the plasma membrane of aerobic bacteria. Although these three membranes differ widely in their composition, the coupling ATPases from each are remarkably similar.

3. The bacterium *H. halobium* has a uniquely simple membrane system containing a single protein species, bacteriorhodopsin, that acts as a light-driven proton pump.

4. The visual pigment, rhodopsin, constitutes up to 90% of the protein of the retinal disc membrane and acts as a light-driven Ca^{++} transporter. The changed Ca^{++} distribution within the rod cell modulates the Na^+ permeability of the plasma membrane, which perturbs the electrical status of the cell and thus alters the pattern of nerve impulses leaving the retinal cells.

5. Synaptic receptors are integral membrane proteins that provide mechanisms for changing the electrical status of the cell membrane in response to chemical messengers or neurotransmitters. The receptor proteins are coupled to ion channels in the post-synaptic membrane. In some cases the channels may be an integral part of the receptor protein, whereas in other systems communication between the receptor and the ion channel may involve a number of membrane-bound enzymes that are modulated by cyclic nucleotides. The opening of these channels may lead to the formation of voltage-dependent ion channels elsewhere in the membrane that allow large ionic currents to flow and thereby generate an action potential.

6. Hormone receptors, with the exception of those for steroid hormones, are integral proteins of the plasma membrane having exposed binding sites at the membrane's outer surface. In many cases the physiological expression of the message carried by the hormone is mediated by intracellular cyclic AMP. In such systems, the hormone receptors interact transiently with the plasma membrane-bound enzyme, adenylate cyclase. The interaction between the hormone receptors and adenylate cyclase may be modulated by the fluidity of the membrane and by a GTP-binding protein/GTPase associated with the cyclase. The overall response of the cell to those hormones whose effects are mediated by cyclic AMP may be subject to further modulation by intracellular Ca^{++} and by intracellular cyclic GMP. Thus hormones that affect the permeability of the plasma membrane to Ca^{++} may modify the effects of cyclic-nucleotide mediated hormones. The binding of a hormone to its receptor may lead to internalization of the receptor-hormone complex. This hormonal regulation of receptors may be a mechanism for the regulation of target-cell sensitivity.

CHAPTER TEN

PHYSICAL METHODS USED IN THE STUDY OF MEMBRANES

IN THIS CHAPTER THE MAJOR PHYSICAL METHODS THAT ARE APPLICABLE TO the study of membranes will be outlined. It is not possible in the space available either to treat fully the theoretical basis of the methods or to give a comprehensive coverage of their applications. Accordingly, the principles of the methods will be dealt with, only in so far as is necessary to indicate the physical significance of the parameters that are measured in membrane studies. The examples of application of the methods are chosen so as to give an idea of the range of information that can be obtained by use of each technique, and to indicate the potential and relative values of the methods.

Nuclear magnetic resonance (n.m.r.)

An atomic nucleus may be regarded as a positively charged spinning ellipsoid and, as such, can have both a magnetic moment and spin angular momentum. When a nucleus with magnetic moment μ is placed in a magnetic field H_0, the magnetic moment will precess about the H_0-direction, inclined to H_0 at a constant angle θ—just as a gyroscope precesses about the direction of the earth's gravitational field (figure 10.1a). The spin angular momentum is quantized and defined by the spin number $I=\frac{1}{2}$, 1, $1\frac{1}{2}$ . . . For nuclei with spin number $I=\frac{1}{2}$ (e.g. ^{1}H, ^{13}C, ^{31}P, ^{19}F), only two ($2I+1$) orientations or *spin states* are possible (figure 10.1a and b). In the lower-energy spin state, the magnetic moment μ is inclined at an angle θ to the positive direction of the field H_0 (figure 10.1a) and, in the higher-energy state (figure 10.1b), the magnetic moment is inclined at the same angle θ to the negative direction of H_0. The energy associated with orientation (a) is $-\mu_z H_0$ and with orientation (b) is $+\mu_z H_0$, where μ_z is the component of μ along H_0. Accordingly, the separation of the two energy levels or spin states is $2\mu_z H_0$, and transitions

238

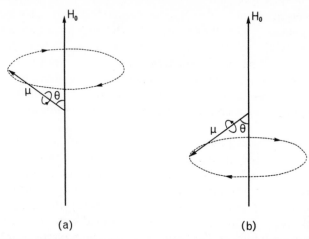

Figure 10.1. Precession of nuclei with spin number $I = \frac{1}{2}$ and magnetic moment μ about the direction of a magnetic field H_0. Two orientations are allowed, one of lower energy (a) and one of higher energy (b), each inclined at an angle θ ($54°$ $44'$) to H_0.

between the two levels can be induced by applied electromagnetic radiation of frequency v, where

$$hv = 2\mu_z H_0 \qquad (1)$$

and h is Planck's constant.

For values of H_0 of the order of 1 tesla (10^4 gauss), which are used in the n.m.r. experiment, v falls in the radiofrequency range. The probabilities of induced transitions between the two energy levels are equal but, because there will initially be a slight excess of nuclei in the lower spin state (figure 10.1a), transition from lower to higher spin states will predominate, giving rise to a net absorption of energy from the applied radiofrequency radiation. This can be detected and recorded as an absorption peak. For nuclei of a given type (i.e. constant μ), only one value of v will cause transitions at a given field strength H_0 (equation 1). Conversely if, as in one possible n.m.r. experiment, v is fixed and the magnetic field H is swept over a suitable range, we might expect all the nuclei of a given type to undergo transitions (*resonate*) when a single value H_0, corresponding to equation (1), is reached. If this were so, no distinction could be made between any of the nuclei of a given type (e.g. protons), and n.m.r. spectrometry would be of little practical value. However, each nucleus is more or less shielded from the applied magnetic field H by its immediate electronic environment, and the field actually experienced at the nucleus (H_{eff}) is not H but $H(1-\sigma)$ where σ is a *shielding constant* characteristic of each individual nucleus in a molecule. Consequently the resonance condition $H_{eff} = H_0$, where H_0 satisfies equation (1), will be met for different values of the applied field H such that $H(1-\sigma) = H_0$. As the applied magnetic field is swept, a spectrum of resonance or absorption peaks will be obtained, and each individual nucleus will have a characteristic position or *chemical shift* in the spectrum.

When specific nuclei have been identified, further information can be obtained from other parameters. The area under a resonance peak, for example, corresponds to the number of chemically identical nuclei giving rise to that peak. Information concerning the mobility of the nucleus can also be obtained from n.m.r. experiments. A nucleus in the higher-energy spin state (figure 10.1b) can revert to its original state (*relax*) by loss of energy to its neighbours. The rate of relaxation is characterized by the *relaxation times* T_1 and T_2, and depends upon the rate of movement of that section of the molecule that contains the nucleus being studied. T_1 and T_2 can be measured directly by sophisticated n.m.r. experiments in which the radiofrequency radiation is applied in pulses rather than continuously. T_2 can also often be obtained directly from the standard n.m.r. experiment in which the line width (width of the absorption peak at half its maximum height) is inversely proportional to T_2.

Nuclei in solid systems generally give rise to very broad absorption peaks. The extent to which a nucleus is shielded from the external magnetic field depends not only on its electronic environment within the molecule, but also upon the orientation of the molecule with respect to the external field, so that many differently-oriented molecules will give an envelope of peaks at slightly different frequencies for any given nucleus. This broadening is known as *chemical shift anisotropy*. A second factor causing peak broadening is the general dependence of relaxation times T_1 and T_2 on the mobility of the nucleus, such that slow motion (e.g. in a solid system) can lead to low relaxation times and to broad peaks (line width $\propto 1/T_2$). Rapid molecular rotation of a molecule in solution will cause averaging of molecular orientation with respect to the external field, so reducing the chemical shift anisotropy, and will also result in high relaxation times. Both factors will lead to narrow absorption peaks and so-called high-resolution spectra. Simple lamellar lipid systems are essentially solid and do not give such spectra and, although valuable information has been obtained from special wide-line n.m.r. techniques, studies on lipid systems usually make use of high-resolution spectra.

These are given by hand-shaken dispersions of lipids in water which contains multilamellar vesicles, and to a greater extent by sonicated single-bilayer vesicles (chapter 6). The exact reason for the sharpness of n.m.r. lines in sonicated vesicles is still a subject of debate. Some workers believe that the fast vesicle rotation is sufficient to explain the narrow lines, whereas others argue that the motions of lipid chains in sonicated vesicles are different from those in unsonicated vesicles (see page 272).

Model lipid systems

The n.m.r. techniques have been most generally applied to studies of protons (^1H) and, although these measurements are technically the simplest, they have proved to be of only limited value in lipid systems. High-resolution *proton n.m.r.* spectra of sonicated dispersions of dipalmitoyl phosphatidylcholine in D_2O at different temperatures are shown in figure 10.2. Some problems of proton n.m.r. are immediately apparent. Lipid molecules contain very many protons with similar chemical environments (e.g. CH_2 protons) which simply give a broad envelope of peaks with similar chemical shifts. In the spectrum at 67°C (figure 10.2a) peaks corresponding to the $-\overset{+}{N}(CH_3)_3$, chain$-CH_2$ and chain$-CH_3$ protons can be distinguished, but separate analysis of individual chain methylene groups is clearly impossible. As the temperature is dropped (figure 10.2b, c, d) a progressive broadening of all the major peaks occurs as the system undergoes a liquid crystalline-gel phase transition (42 °C), and the mobility of the protons in the bilayer (particularly $\overset{+}{-}CH_2-$ and $-CH_3$) is reduced. Although a plot of linewidth of the $\overset{+}{N}(CH_3)_3$ protons against temperature shows a sharp upward inflexion as the temperature decreases through the transition point, it is evident from figure 10.2d that the $\overset{+}{N}(CH_3)_3$ group retains considerable mobility, even when the aliphatic chains are frozen in the gel phase.

The lack of resolution is a major (but not the only) problem in the application of proton n.m.r. to lipid systems, and studies of other nuclei have generally proved to be more fruitful. ^{13}C n.m.r. spectra tend to be simpler, having larger chemical shifts and narrower lines, and they can be observed from the natural ^{13}C content of lipids if the sensitivity of the n.m.r. spectrometer is enhanced by computer accumulation of repeated spectral scans. The scan time is reduced in practice by simultaneously exciting nuclei over the whole spectral range, using a short intense radiofrequency pulse and converting the information obtained into a conventional n.m.r. spectrum by a *Fourier transform* operation carried

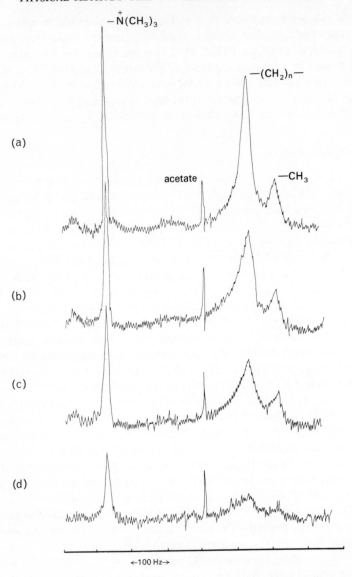

Figure 10.2. High-resolution proton n.m.r. spectra of sonicated dispersions of dipalmitoyl phosphatidylcholine in D_2O at different temperatures. (*a*) 67 °C; (*b*) 52 °C; (*c*) 42 °C; (*d*) 39 °C: from A. G. Lee *et al.*, *Biochim. Biophys. Acta* (1972), **255,** 43–56. The acetate peak arises from the use of acetate-containing buffers.

out by computer. A ^{13}C n.m.r. spectrum of a sonicated dispersion of dipalmitoyl phosphatidylcholine at 64 °C is shown in figure 10.3, comparison of which with figure 10.2 illustrates the higher resolution obtainable with ^{13}C n.m.r. Not only are the chain carbons 2, 3, 14, 15 and 16 all clearly resolved in the ^{13}C spectrum, but the glycerol and choline carbons can also be identified by comparison with standard

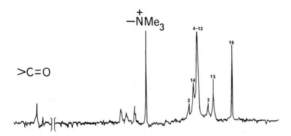

Figure 10.3. ^{13}C n.m.r. spectrum (proton decoupled) of a sonicated aqueous dispersion of dipalmitoyl phosphatidylcholine at 64°C: from Levine *et al.*, *Biochemistry* (1972). **11**, 1416–1421.

compounds. The effect of protons on the ^{13}C spectrum has been decreased by irradiation of the protons at their resonance frequency with a strong radiofrequency field, which causes them to fluctuate rapidly between spin states, so averaging out their coupling effect. This technique, known as *proton decoupling*, is commonly applied in producing n.m.r. spectra of nuclei other than protons.

Using Fourier transform techniques the ^{13}C relaxation times T_1 can be measured for all carbons that give resolved peaks in the n.m.r. spectrum. Moreover, the peak from a specific carbon atom in the middle of the aliphatic chain (e.g. carbon 7) can be obtained by using lipid molecules synthetically enriched with ^{13}C at that position. Detailed analysis of T_1 values obtained in these ways from dimyristoyl phosphatidylcholine above its phase transition temperature have demonstrated the presence of equal rotational motion about successive carbon bonds up to carbon 7 while, at the terminal methyl end of the chain, motion is much faster. More elaborate experiments have suggested that the fatty acyl chains of phospholipids are subject to a 'whipping' motion which can bring the end of the chain up towards the polar surface of the lipid bilayer.

^{2}H n.m.r. of deuterium-labelled phospholipids is ideally suited to the measurement of anisotropic motion of the fatty acyl chains. Deuterium differs from protons and ^{13}C ($I = \frac{1}{2}$) in that its spin quantum number $I = 1$. The ^2H nucleus accordingly possesses an electric moment which interacts with the local electric field gradient. This *quadrupole* interaction changes the absorption frequency of the deuterium nucleus and depends upon the orientation of the molecule with respect to the applied magnetic field. For a rapid *isotropic* motion (i.e. with equal probabilities in all directions) of the nucleus, these effects will average out and the ^2H n.m.r. spectrum will consist of a single line, whereas if the motion is anisotropic then the spectrum will appear as a doublet. Although in a lipid bilayer the rotation about C-C bonds can be quite fast, this motion is not isotropic, because the long axis of the fatty acyl chains has a preferred orientation in space (i.e. perpendicular to the bilayer surface). The chain motion does not, therefore, completely eliminate the doublet line splitting, but only reduces it and, from the extent of the residual splitting, it is possible to estimate the range of angles through which any given deuterium atom can move. This range of movement is expressed as an *order parameter* (S_{mol}) where lower values of S_{mol} indicate greater motional freedom. Figure 10.4 shows the order parameters for different deuteriated carbon atoms in the palmitoyl chains of unsonicated dispersions of 1,2-dipalmitoyl-*sn*-glycero-3-phosphocholine and of 2-oleoyl-1-palmitoyl-*sn*-glycero-3-phosphocholine. The order parameters can be seen to be fairly constant for the first nine chain segments, but then to decrease rapidly in the central part of the bilayer in a pattern of mobility similar to that predicted by analysis of ^{13}C n.m.r. T_1 values (see above).

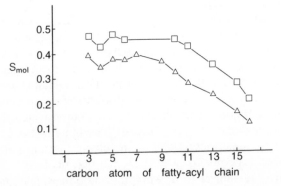

Figure 10.4. Order parameter S_{mol} as a function of the labelled carbon atom of the fatty acyl chain for dipalmitoyl phosphatidylcholine (1,2-dipalmitoyl-*sn*-glycero-3-phospho-choline) (□) and for 2-oleoyl-1-palmitoyl-*sn*-glycero-3-phosphocholine (△) at 42 °C: from Seelig and Seelig, *Biochemistry* (1977), **16**, 45–50.

It is interesting that the incorporation of an unsaturated fatty acyl chain causes a general decrease in the order of the bilayer.

^{2}H n.m.r. has also been applied to the study of the anisotropic motion of CD_2 groupings in the head groups of multilamellar dispersions of deuteriated phospholipids. Analysis similar to that described above for fatty acyl chains led to the conclusion that the head groups of phosphatidylcholine and phosphatidylethanolamine lie essentially parallel to the surfaces of their respective bilayers and rotate rapidly about an axis perpendicular to the bilayer.

Unlike ^{2}H (but like ^{1}H and ^{13}C), ^{31}P has spin quantum number $I = \frac{1}{2}$ and so does not show quadrupole splitting. ^{31}P n.m.r. signals from molecules with restricted motion, as in unsonicated lipid bilayers, are however broadened by chemical shift anisotropy (page 240) and analysis of the line shape can give information about the limited mobility of the ^{31}P nucleus. Comparison of ^{31}P n.m.r. spectra from the phosphate groups of unsonicated phospholipid bilayers with spectra calculated assuming certain rotational modes have led to patterns of head group mobility largely consistent with those derived from ^{2}H n.m.r. data.

Limited use has been made of a fifth nucleus, ^{19}F ($I = \frac{1}{2}$) as a reporter group in lipid bilayer studies. Molecules of the type $CH_3(CH_2)_m CHF$ $(CH_2)_{n-2}COOH$ have been incorporated into sonicated phosphatidylcholine vesicles and the ^{19}F resonances used to report on molecular movement at various points along the fatty acid chain. Values of T_2 (page 240) were determined from the widths of the ^{19}F peaks and were found to increase as the F atom moved nearer to the methyl end of the chain, consistent with increasing motion in this direction. More recently, *gem* difluorinated myristic acids have been used to follow complex phase transitions in dimyristoyl phosphatidylcholine-distearoyl phosphatidylcholine dispersions and in lipids extracted from *E. coli* membranes.

Information about lipid systems can also be obtained from the interactions of component n.m.r. signals with suitably located reference groups [e.g. spin labels (page 249) or paramagnetic ions]. Paramagnetic ions shift or broaden n.m.r. signals and an early experiment of this type used Mn^{++} (which causes line broadening of signals from immediately-neighbouring protons) to study sonicated lipid vesicles. Addition of Mn^{++} to a sonicated lipid-water system reduced the $-\overset{+}{N}(CH_3)_3$ proton peak (cf. figure 10.2) to 40% of its original intensity, corresponding to broadening out of the signals from those quaternary ammonium protons (approximately 60% of the total) located in the outer monolayer of the vesicle. Sonication of the lipid in the presence of Mn^{++} allowed the metal ions to occur both inside and outside the vesicles, when all the $-\overset{+}{N}(CH_3)_3$ signals disappeared. The conclusions drawn from these experiments were not only that the sonicated vesicles were bounded by a single bilayer, but also that they were effectively sealed (chapter 6). This approach has since been developed to estimate the sizes of lipid vesicles using ^{31}P n.m.r. In contrast to multilamellar liposomes (page 104), the smaller sonicated lipid vesicles tumble fast enough to average out the chemical shift anisotropy of ^{31}P signals which appear as sharp lines. Cationic shift and broadening agents have been used to obtain the 'outside-inside' ratio, and hence the sizes of such vesicles (16–35 nm), and to correlate sizes with hydrocarbon chain length and cholesterol content of the lipid constituents.

Biological membranes

Despite extensive applications of n.m.r. techniques to model lipid systems, relatively little information has been derived by these means from biological membranes. The presence of a range of lipids, including cholesterol, together with protein, leads to overlapping signals from molecules in many different environments, and at best only average information can be obtained from normal component nuclei. A ^{13}C n.m.r. spectrum of mitochondrial membrane (which shows better resolution than most natural membranes) is illustrated in figure 10.5.

There are relatively few ^{31}P nuclei in natural lipid systems, and this is an advantage in ^{31}P n.m.r. studies of biological membranes. ^{31}P line shapes derived from a number of natural membranes (including erythrocytes, chromaffin granule membranes, viral and microbial membranes) have been analysed and shown to be consistent with the presence of liquid crystalline phospholipids undergoing restricted anisotropic motion in

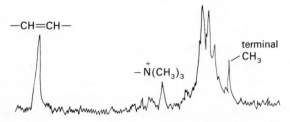

Figure 10.5. High-resolution ^{13}C n.m.r. spectrum at 25·5 MHz at 37°C of unsonicated mitochondrial membrane: from Keough *et al.*, *Chem. Phys. Lipids* (1973), **10**, 37–50.

large bilayer structures. In contrast, a ^{31}P n.m.r. study of bovine and rat liver microsomal membranes showed that a significant portion of the constituent phospholipids experience isotropic motion, and it was suggested that this might reflect 'flip-flop' bilayer transfer of phospholipids (page 123) in these biosynthetically-active membranes.

Incorporation of probe nuclei into biological membranes has also given some results. Thus membranes from the yeast *Candida utilis* grown on ^{13}C-enriched acetate have been shown to give good ^{13}C n.m.r. spectra, while preliminary experiments have demonstrated that [methyl-^{13}C] choline can be incorporated into the membranes of Chinese hamster ovary cells in culture. Deuterium has also been introduced into natural membranes and biosynthetically incorporated [15-^{2}H$_2$]-palmitic acid has been used to follow phase transitions in the membrane of *Acholeplasma laidlawii* (page 272). This type of approach is relatively recent, and an assessment of its general utility is not yet possible.

Electron spin resonance (e.s.r.)

The nitroxide free radical ($>$N—Ȯ), by virtue of its unpaired electron spin, gives rise to an electron spin resonance signal which is sensitive both to the detailed motion of the radical and to the polarity of its environment. If a molecule containing this radical is incorporated into a bilayer or membrane, therefore, information can be obtained concerning its surroundings. The technique has the advantage that only the signal from the probe itself need be considered, and the disadvantage that insertion of the probe could disturb the molecular relationships of the system it is intended to describe.

Like an atomic nucleus with spin number $\frac{1}{2}$, the unpaired electron has both a magnetic moment and a spin angular momentum, and when placed in a magnetic field can adopt only two orientations relative to the field (cf. figure 10.1 but note that the particle spin-magnetic moment relationship is reversed). As in the case of the nucleus, the energy difference between the two orientations is proportional to both the magnetic moment and the field strength, and transitions between the two levels can be induced by electromagnetic radiation of suitable frequency. For field strengths of 0·34 tesla (3400 gauss) such as are commonly used in e.s.r. equipment, v is of the order of 10^{10} Hz, which is microwave frequency. At resonance, microwave radiation is absorbed and recorded as an absorption peak, or more commonly, as its first derivative (figure 10.6).

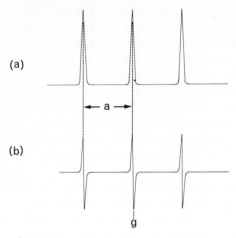

(a)

(b)

Figure 10.6. Freely tumbling nitroxide radical e.s.r. spectrum shown as (a) absorption peaks and (b) their first derivatives.

The nitrogen nucleus ^{14}N has a spin number $I = 1$ and, as such, can itself assume three $(2I+1)$ possible orientations in the magnetic field. Each orientation of the nitrogen nucleus will correspond to a slightly different magnetic environment for the unpaired electron, which accordingly shows not one, but three, resonances as the magnetic field is swept and v is held constant. The oxygen nucleus ^{16}O is non-magnetic with spin number $I=0$ and does not further split the three lines. The e.s.r. spectrum of a freely-tumbling nitroxide radical is shown in figure 10.6, both as absorption peaks and as their corresponding first derivatives. Both the g-value, which specifies the position of the middle peak in the spectrum as the field is swept, and the *hyperfine splitting* (the line separation) are sensitive to the environment and orientation of the nitroxide radical.

The nitroxide radical is a resonance hybrid of the structures:

$$\left[\begin{array}{ccc} \diagdown \diagup & & \diagdown \diagup \\ N & \leftrightarrow & N^{+} \\ | & & | \\ O^{\cdot} & & O^{-} \end{array} \right]$$

and the contribution of the charged form increases with the polarity of the medium. Thus transfer of the radical from a non-polar to a polar environment results in a higher spin density on the nitrogen atom, and the hyperfine splitting increases. The g-value is rather less sensitive to medium polarity and decreases as the polarity increases. When a radical, or *spin label*, partitions between an aqueous and a hydrocarbon environment, two superposed spectra can be seen, provided that the exchange between the two media is relatively slow. Partition of the spin label TEMPO (figure 10.7) between water and dispersed dielaidoyl phospha-

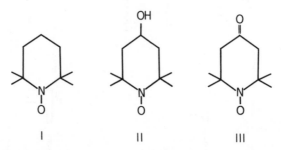

Figure 10.7. Some simple nitroxide spin-labels: I, TEMPO (2,2,6,6-tetramethylpiperidine-1-oxyl); II, TEMPOL (2,2,6,6-tetramethylpiperidine-1-oxyl-4-ol); III, TEMPONE (2,2,6,6-tetramethyl-4-piperidinone-1-oxyl).

tidylcholine (elaidoyl = *trans* 18:1) above its transition temperature is shown in figure 10.8a. In the hydrophobic medium g is increased relative to water, and the spectrum shifts to low field (to the left) while the hyperfine splitting is decreased, with the result that maximum separation between the two superposed spectra is seen in the high field lines. Thus the signal H (figure 10.8a) arises from TEMPO in the hydrophobic lipid and signal P from the label in water. The amplitudes of the two peaks reflect the partition of the probe between the two environments so that $f[=H/(H+P)]$ is the fraction of the spin label dissolved in the lipid phase. This fraction can be seen (figure 10.8) to decrease (i.e. H decreases) as the lipid is cooled through its liquid crystalline-gel phase transition temperature, and plots of f versus temperature have been used to demonstrate phase transitions for a number of phospholipids (figure 10.9). Whereas the main phase transition of a single phospholipid is usually sharp (figure 10.9), solid and fluid phases in a binary lipid mixture can coexist over a considerable temperature range, the beginning and end of which can be detected as breaks in f versus temperature plots. Such data derived from a series of binary mixtures of different composition can then be used to plot a phase diagram, and a number of phospholipid pairs have been analysed in this way.

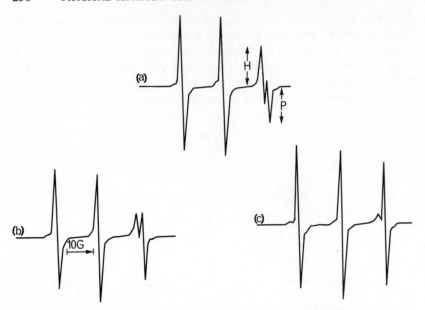

Figure 10.8. E.s.r. spectra of the spin label TEMPO in an aqueous dispersion of the phospholipid, dielaidoyl phosphatidylcholine, above (a), at (b) and below (c) the lipid phase transition temperature. H and P are the high field signals in the hydrophobic lipid and polar aqueous environments respectively: from S. H. W. Wu and H. M. McConnell, *Biochemistry* (1974), **14,** 847–854.

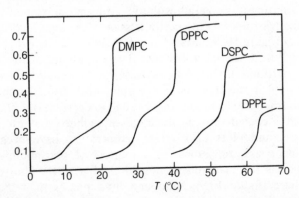

Figure 10.9. Temperature variation of the spectral parameter f for aqueous dispersions of dimyristoyl phosphatidylcholine (DMPC); dipalmitoyl phosphatidylcholine (DPPC); distearoyl phosphatidylcholine (DSPC) and dipalmitoyl phosphatidylethanolamine (DPPE). Note the broad 'pretransition' in DMPC, DPPC, and DSPC which is confined to phosphatidylcholines (see p. 268) from E. J. Shimshick and H. M. McConnell, *Biochemistry* (1973), **12,** 2351–2360.

The *g*-value and hyperfine splitting of a nitroxide e.s.r. spectrum depend upon the orientation of the spin label, and a different three-line spectrum results from each of three principal mutually-perpendicular orientations of the nitroxide radical with respect to the magnetic field. The spectral parameters corresponding to each fixed orientation can be determined from single crystals of a host substance which contain low concentrations of uniformly orientated spin labels, and these parameters can be used to calculate spectra for labels in intermediate orientations. When a spin-label is undergoing rapid *isotropic* tumbling (page 244), the directional effects (anisotropies) of the *g*-values and hyperfine splittings are averaged, and a sharp three-line spectrum (figure 10.6) results. Alternatively, an immobilized random array of spin labels gives a *powder*

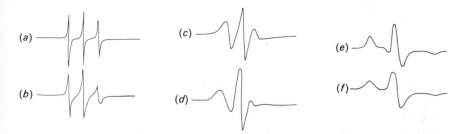

Figure 10.10. The effect on the e.s.r. spectrum of progressively slowing down an isotropically-tumbling spin label. The spectra are given by the spin label **TEMPO** in glycerol. As the temperature falls, the viscosity of the medium increases, and the motion of the spin label decreases. (*a*) 43°C; (*b*) 26°C; (*c*) 9°C; (*d*) 0°C; (*e*) −36°C; (*f*) −100°C: from P. Jost *et al.* in *Structure and Function of Biological Membranes* (1971), (ed. L. I. Rothfield), New York, Academic Press, pp. 83–144.

spectrum, which is the sum of individual spectra for molecules at all possible orientations and which can be predicted from a knowledge of the parameters for each of the three principal orientations. The effect of progressively slowing down an isotropically-tumbling spin-label is shown in figure 10.10. It is clear from figure 10.10 that the shapes of the spectra to some extent reflect the mobilities of the spin label, and it is generally true that, as in the case of n.m.r., wide lines are indicative of slow motion.

Membrane components are not, in general, free to undergo isotropic motion and fatty acyl chains, for example, are more or less constrained to move anisotropically about the direction of the normal to the bilayer

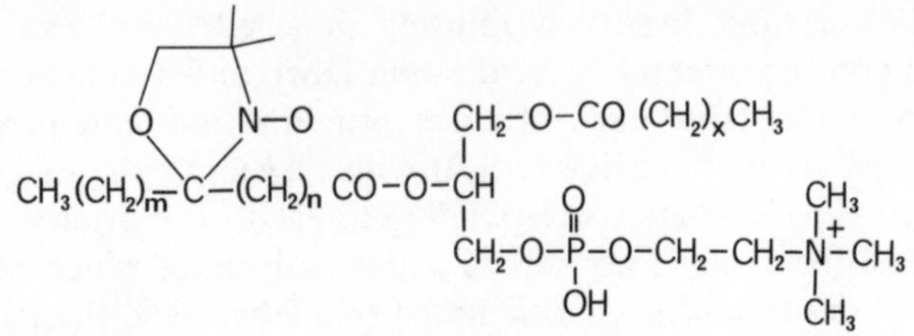

Figure 10.11. Phosphatidylcholine IV containing a spin label, the 2,2-dimethyl-N-oxyl-oxazolidine (doxyl) moiety in its C-2 fatty acyl chain.

surface. Spin-labelled fatty acids and phospholipids (e.g. IV, figure 10.11) have been used to report on the detailed motion of fatty acyl chains in orientated bilayers. Such motion is usually discussed in terms of an *order parameter*, analogous to that derived from n.m.r. data, which is a measure of the distribution of molecular orientations relative to the normal to the membrane surface. The order parameter can be derived from spectra of specific orientations of the bilayer with respect to the applied magnetic field, and from a knowledge of the spectral parameters of the principal orientations of the nitroxide radical itself. As the amplitude of motion of a flexible chain increases, the order parameter decreases, so that the plot shown in figure 10.12*b* represents an increasing mobility of the spin label as it moved down the fatty acyl chain (i.e. n increases) away from the head group of the phosphatidylcholine IV (figure 10.11). This *flexibility gradient* is a common feature in the fatty acyl chains of membrane components, and a casual comparison of the spectrum in figure 10.12*a* (iv) with those in figure 10.10 will show that the mobility of a spin label near the methyl end of the fatty acid chain, in at least this case, is comparable with that of TEMPO in glycerol at or above room temperature. Order parameters for phospholipid bilayers have also been derived from ^2H n.m.r. data (page 244) and considerable debate has arisen from the fact that, although both methods indicate the presence of a flexibility gradient down the fatty acyl chains in comparable bilayers, the details of the gradient are quite different (figures 10.4 and 10.12*b*) depending on the technique used. These discrepancies may arise from the different time scales of ^2H n.m.r. and e.s.r., and recalculation of the data of figure 10.12 can, in fact, lead to a flexibility gradient quite similar to that of figure 10.4. A more fundamental point, however, is that spin labels in general, and the relatively polar doxyl group (figure 10.11) in particular, might be expected to perturb the order of the lipid bilayer more than do simple deuteriated lipids. This is clearly always a problem in probe techniques and needs to be at least considered in evaluation of data from e.s.r. experiments.

Spin labels have been used to measure diffusion coefficients for lateral diffusion of lipids in bilayers. At high concentrations of spin-labelled molecules, the e.s.r. line shapes change as a result of the interaction between neighbouring spin labels, and an empirical relationship can be established between line shape and label concentration. In one experiment the labelled phosphatidylcholine analogue V (figure 10.13) was introduced as a small patch into a planar array of unlabelled hydrated egg phosphatidylcholine, and the fall-off in local concentration caused by lateral diffusion was followed by e.s.r. line shape analysis. This led to a diffusion coefficient $D = 1 \cdot 8 - 10^{-8}$ cm^2/s at 25 °C. Similar approaches have been used to monitor lateral diffusion of other spin-labelled lipids such as IV (figure 10.11) and VI (figure 10.13) in a range of artificial and natural (e.g. sarcoplasmic reticulum) membranes.

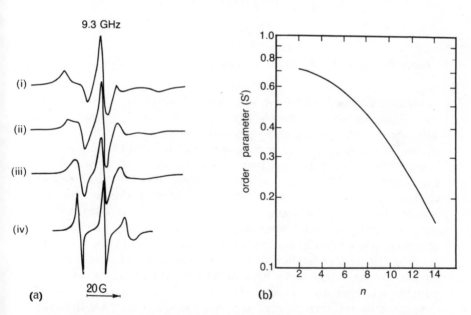

Figure 10.12. (a) E.s.r. spectra of the spin-labelled phosphatidylcholine IV (figure 10.11) in oriented egg phosphatidylcholine bilayers for varying values of m,n. (i) $m=10$, $n=3$, (ii) $m=7$, $n=6$, (iii) $m=5$, $n=10$, (iv) $m=1$, $n=14$. (b) Plot of order parameters, S', derived from the spectra in (a), versus the distance of the spin label from the polar head group of the lipid: from B. J. Gaffney and H. M. McConnell, *J. Magn. Res.* (1974), **16**, 1–28.

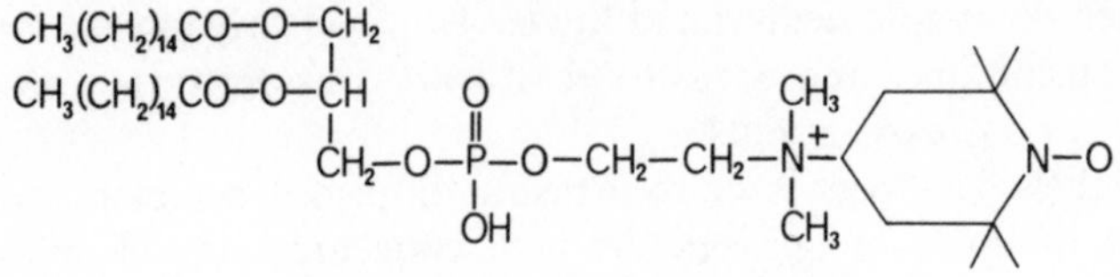

Figure 10.13. Spin-labelled analogues of phosphatidylcholine (V) and cholesterol (VI).

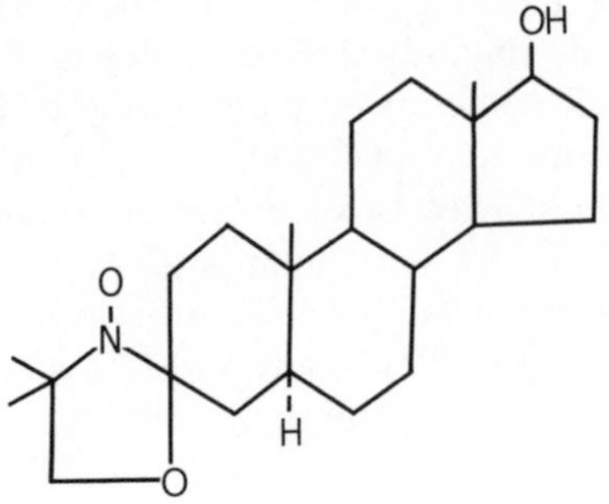

Transmembrane lipid motion (flip-flop) has also been followed using e.s.r. The spin-labelled phosphatidylcholine analogue V (figure 10.13) was incorporated into egg phosphatidylcholine vesicles, and those labels situated in the outer monolayer were selectively reduced by ascorbic acid which destroys the paramagnetism (and hence the e.s.r. signal) of the nitroxide radical. The vesicles were impermeable to ascorbic acid, so that the labels of the inner lipid monolayer were unaffected. The ascorbic acid was removed by gel filtration, and the asymmetrically-labelled vesicles were incubated at 37 °C, when any labels subsequently reaching the monolayer via a flip-flop mechanism could de detected by ascorbic acid titration. The rate of flip-flop determined by these means was found to be slow, as has subsequently been found generally to be the case with pure phospholipid vesicles (page 124).

Much of the information derived from e.s.r. is also available from the application of other physical methods outlined in the present chapter. Spin-label studies have been particularly useful, however, in studying the anisotropic motion of membrane components, and have the advantage over n.m.r. of much higher sensitivity and (in the case of ^{2}H n.m.r.) of more readily-available probe molecules. The major disadvantage of spin labels is, as discussed above, the danger that the probe itself might affect the system being studied, but effort is currently being expended in the development of probe molecules designed to disturb membranes to a minimal extent, and their application will be of considerable interest.

Fluorescence

All molecules have a set of electronic energy levels (S_0, S_1 and T_1, figure 10.14) within each of which vibration levels provide further subdivisions. At room temperature, most molecules are in the lowest vibrational level of the ground electronic state S_0. Absorption of ultraviolet or visible light of frequency v leads to absorption of energy E ($= hv$, where $h =$ Planck's constant) and the molecule is excited to the excited singlet state S_1, or to a higher state. The energy level to which the molecule is raised depends on the frequency of the absorbed radiation. After excitation, molecules in solution very rapidly revert to the lowest vibrational level of S_1 without emission of radiation, after which they can return to the ground state by a number of pathways. Direct demotion to an inter-

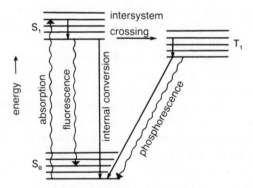

Figure 10.14. Electronic transitions leading to fluorescence and competing processes: S_0 = ground electronic state; S_1 = lowest excited singlet state; T_1 = triplet state. In the singlet states all electrons have paired spins, whereas in a triplet state two electron spins are unpaired. Wavy arrows denote transitions with absorption or emission of radiation. Straight arrows denote non-radiative transitions.

mediate vibrational state of S_0 can occur with emission of light, and this is the process known as *fluorescence*. Because the energy change is less than that involved in the initial absorption (figure 10.14), the frequency of light emitted in fluorescence will be lower than that of the incident light, i.e. *fluorescence is at longer wavelength than the absorption band*, and irradiation with ultraviolet light commonly leads to fluorescence in the visible region. Non-radiative demotion to the ground state S_0 (*internal conversion*) and *intersystem crossing* to the triplet state T_1 (figure 10.14) compete with fluorescence. Reversion from T_1 to the ground state occurs mainly by non-radiative emission for molecules in solution at room temperature when radiative emission (phosphorescence) does not successfully compete. In addition to these mechanisms, intermolecular interactions of the excited molecule with another (quenching) species can also act to decrease the efficiency of fluorescence, which is measured as the *quantum yield* (the fraction of the absorbed light that is re-emitted as fluorescence).

The electronic and vibrational energy levels are affected by the polarity of a molecule's immediate environment. The dipole moments in the different electronic states are often in the order $S_1 > T_1 > S_0$ in which case transfer of the molecule from a polar to a non-polar environment will tend to cause destabilization of S_1 relative to S_0 and to T_1. Increase in the S_1-S_0 energy gap will lead to a shift of fluorescence maxima to higher frequency (i.e. shorter wavelength, a 'blue shift') while increase in the S_1-T_1 gap can reduce the probability of intersystem crossing and so give increased quantum yields. A suitable fluorescent molecule can accordingly be used to probe the polarity of a membrane.

The fluorescent probes are not only sensitive to the polarity of their environment, but also to the viscosity. A probe molecule which has been newly promoted to its excited state can be stabilized by rearrangement of solvent molecules around it. If the viscosity of the medium is increased, this stabilization is inhibited, resulting in an increased S_1-S_0 energy gap and a blue shift in fluorescence. This blue shift is, as in the case of that induced by polarity changes, often accompanied by an increase in the quantum yield of fluorescence.

The mobility of a fluorescent probe can also be assessed and used to report on the viscosity of its immediate environment. Excitation of the probe (*fluorophor*) with plane-polarized light will lead to the emission of light polarized in the same plane, except in so far as the emitter has rotated during the lifetime of its excited state. The *fluorescence polarization P* and the *fluorescence anisotropy r* are defined by the equations

$$P = \frac{I_{\parallel} - I_{\perp}}{I_{\parallel} + I_{\perp}} \; ; \quad r = \frac{I_{\parallel} - I_{\perp}}{I_{\parallel} + 2I_{\perp}}$$

where $I_{\parallel}$ and $I_{\perp}$ are the fluorescence intensities parallel and perpendicular to the plane of polarization of the excitation beam. The parameters P and r reflect the mobility of the fluorophor and can be used to obtain a value for the microviscosity of its environment.

A variety of fluorescent probes has been used to study the fluidity of lipid bilayers and biological membranes. Most studies of this type have, however, employed 1,6-diphenyl-1,3,5-hexatriene (DPH) and perylene (figure 10.15). DPH and perylene form aqueous dispersions that are almost devoid of fluorescence but, in the presence of liposomes, the fluorescence signal increases sharply (up to 1000-fold) as the probes are incorporated into the non-polar lipid bilayers. The probes partition equally well into the liquid crystalline and gel phases of the lipid, and fluorescence polarization measurements have been extensively used to determine the average microviscosity of the lipid phase. Measurements of this type can monitor phase transitions in liposomes and have given information about the effects of various components [e.g. cholesterol, unsaturated fatty acyl chains, sphingomyelin (see chapter 6)] on the fluidity of the lipid bilayer. DPH particularly has been used to study the fluidity characteristics of a wide range of isolated biological membranes, and even of membranes in intact cells. In the latter case, account must be taken of the fact that the probe is not necessarily confined to the cell

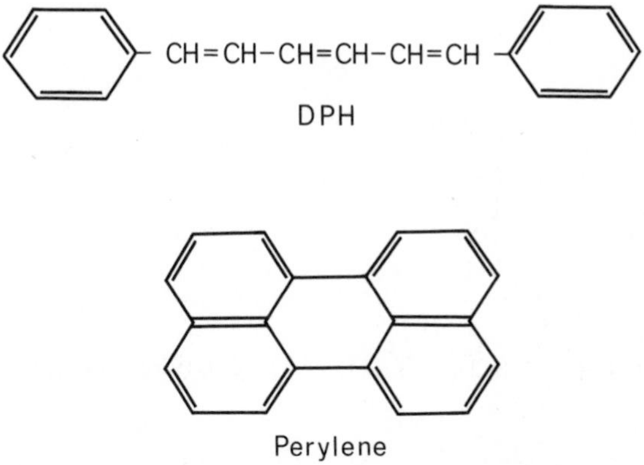

DPH

Perylene

Figure 10.15. Fluorescent probes DPH (1,6-diphenyl-1,3,5-hexatriene) and perylene.

surface membrane, and can partition into all the membranes of the cell. Nevertheless, the method offers an opportunity to follow membrane fluidity changes associated with different physiological states of the cell, such as differentiation or transformation.

DPH and perylene are distributed throughout the lipid phase of a membrane and are accordingly not suited to report on defined regions of the bilayer. Fluorescent molecules are available which can be used in this way, however. ANS (figure 10.16) binds to the head group region of membranes with concomitant enhancement and blue shift of the fluorescence signal, whereas the anthranoyl probes AS and AP (figure 10.16) are believed to sample the hydrocarbon interior of lipid bilayers approximately 1·5 nm and 0·5 nm respectively inwards from the phosphate head groups.

Compounds like AS and AP which contain fluorescent dyes covalently attached to fatty acids or phospholipids have been used in fluorescent polarization studies but, in general, the partial restriction of rotational

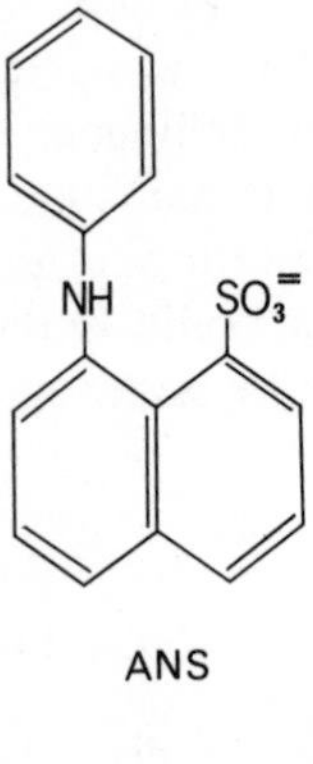

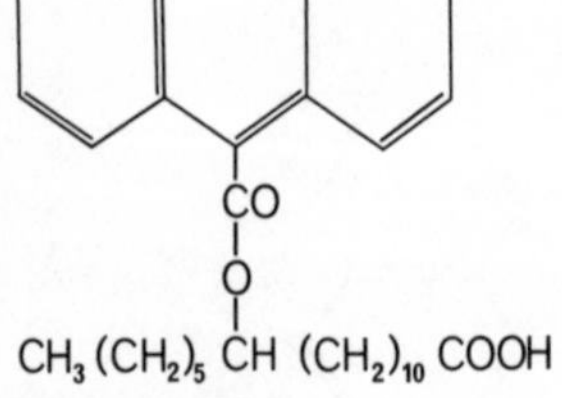

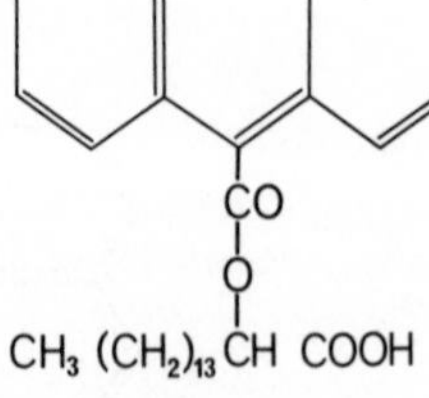

Figure 10.16. Fluorescent probes ANS (1-anilinonaphthalene-8-sulphonate), AS 12-(9-anthranoyl) -stearic acid and AP 2- (9-anthranoyl) -palmitic acid.

modes in such compounds complicates correlations of fluorescence polarization parameters with microviscosity. ANS, AS and AP have been employed to detect structural transitions in the mitochondrial membrane as a result of electron transport. All three compounds, probing the lipid bilayer at different depths, show fluorescence changes which are associated with electron transport. The time constants for the

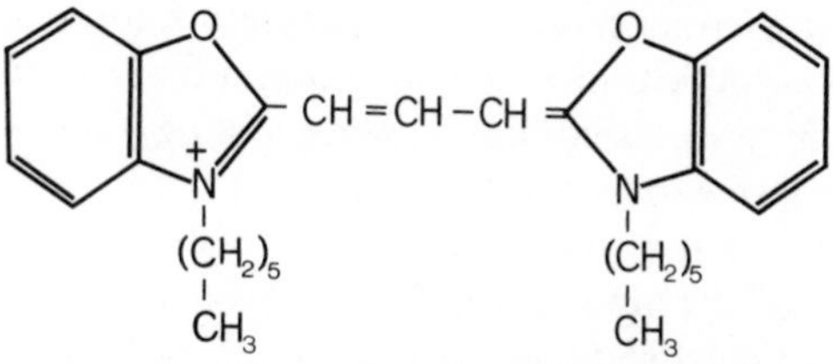

Figure 10.17. The fluorescent probe 3,3′-dihexyloxacarbocyanine.

changes detected by ANS, AP and AS decrease in that order, i.e. as the probe location moves further from the head groups, giving an overall picture of rapid electron transport-linked changes in the centre of the mitochondrial membrane which slow down as the interface is approached. The mechanisms of these fluorescence changes are not fully understood, but experiments of this type are potentially of great value in following rapid functional changes in membranes. A further class of fluorophors has recently been introduced which show large changes in fluorescence intensity in response to changes in membrane potential. Thus the fluorescence of 3,3′-dihexyloxacarbocyanine (figure 10.17) incorporated into red blood cell membranes has been shown to increase by up to 50% following valinomycin-induced hyperpolarization of the membrane, suggesting the possible general application of such compounds in measurements of membrane potentials.

All the fluorescent probes so far discussed are designed to report on the lipid phase of membranes, but membrane proteins also can be studied by these means. Indeed, most proteins carry intrinsic fluorescent probes in the form of tryptophan residues, the fluorescence characteristics of which are determined by their immediate environment. The effect of a conformational change of a functional protein on its fluorescence is well illustrated by the binding of cholera toxin to G_{M1} ganglioside (chapter 7). Cholera toxin shows a tryptophan fluorescence which is enhanced and shifted to shorter wavelengths on specific interaction with G_{M1}, indicating a conformational change in the protein which results in transfer of the tryptophan group to a less polar environment. It may be that this conformational change occurs on binding to the target membrane as a prerequisite to formation of an active A sub-unit (chapter 7). Although involved in membrane interaction, cholera toxin is not, of course, a membrane protein, but conformational changes of membrane proteins ensuing from binding of specific ligands have been followed by fluorescence methods. Thus the Ca^{++}-ATPase from sarcoplasmic reticulum (page 183) showed enhancement of tryptophan fluorescence on binding of Ca^{++}, and the acetylcholine receptor protein from *Torpedo marmorata* (page 217) undergoes rapid fluorescence changes on binding of the agonists acetylcholine or suberylcholine.

The lateral diffusion of membrane components, particularly proteins, within the plane of the lipid bilayer is of great current interest in view of the possibilities of constraint imposed upon otherwise mobile proteins by cytoskeletal and even extracellular elements (chapter 6). The application of spin-labelling to the measurement of lateral diffusion of membrane lipids has already been mentioned, and a number of fluorescence methods have recently been developed to complement and extend these studies. *Fluorescence photobleaching recovery* involves labelling membrane proteins with a fluorophor and bleaching a small area of the labelled membrane with a high-intensity spot of light. The bleaching decreases the fluorescence of the irradiated area, and the signal will increase only as a result of diffusion of unbleached molecules back into the area of study. The rate of recovery of fluorescence can accordingly be used as a measure of the rate of lateral diffusion of labelled protein. Fluorescence photobleaching is suitable for measurement of diffusion coefficients (D) of less than 10^{-9} cm^2/s which is relevant to most integral membrane proteins so far studied. Membrane phospholipids show faster diffusion rates, and these can be followed by *fluorescence correlation spectroscopy*. In this technique, a laser beam continuously excites labelled molecules within the area of a small circle (3 μm radius) on a lipid bilayer surface. Only those molecules within the irradiated area will fluoresce, and the intensity of the signal will fluctuate as lateral diffusion causes variation in the number of fluorophors in the target area. Analysis of fluorescence intensity fluctuations with time can accordingly give a value for the diffusion coefficients of labelled molecules. Both fluorescence photobleaching recovery and fluorescence correlation spectroscopy have been used to obtain the diffusion coefficients of membrane components quoted in chapter 6 (page 117).

One further fluorescence technique can give information about the mobility of membrane components and their mutual interactions. *Resonance energy transfer* can occur between two different fluorophors, provided that the emission spectrum of the donor overlaps the absorption spectrum of the acceptor, and provided that the interacting species are less than 10 nm apart. In practice, excitation of one fluorophor (the donor) by light at its absorption maximum will lead not only to fluorescence at the characteristic emission wavelength of the donor but also at that of the acceptor as energy is transferred. The fluorescence intensity of the donor will also be decreased relative to its normal value, and the ratio of the fluorescence intensities (receptor/donor) will give a measure of the extent of interaction. Hence measurement of the fluorescence ratio can be used to follow the association of differentially-labelled membrane integral proteins, because resonance energy transfer between two species is increased as they come together. This approach has been used to follow the aggregation of concanavalin A receptors in normal and transformed fibroblasts by employing two populations of concanavalin A, one labelled with rhodamine (the acceptor) and one labelled with a dansyl group (the donor).

Diffraction methods

X-ray diffraction

Stacked arrays of parallel membranes or polar lipid bilayers behave like an optical diffraction grating in forming diffraction patterns from a beam of X-rays. The X-rays are randomly scattered from the atoms in the bilayers, but it can be shown that the conditions for constructive interference of scattered radiation can be derived by considering the incident X-rays as being reflected by successive planes in the array (figure 10.18). Constructive interference will occur between radiation 'reflected' from equivalent points A and B in adjacent bilayers, provided that the path length difference (CB+BD, figure 10.18) is an integral number of wavelengths. For radiation incident at an angle θ to the plane of the bilayers, this condition, known as Bragg's law, may be written as:

$$2d \sin \theta = n\lambda$$

where d (figure 10.18) is the lamellar spacing, λ is the wavelength of the X-rays, and n is an integer.

Constructive interference will give a diffracted beam which may be detected on a photographic film. The Bragg condition will be met for a

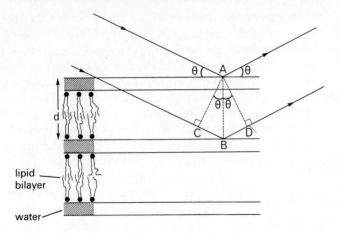

Figure 10.18. Diffraction of an X-ray beam from stacked polar lipid bilayers. Hydrated phospholipids contain water in the hydrophobic head group regions separating the bilayers. The lamellar spacing d, corresponding to the repeat distance in the array, includes both the lipid bilayer and the water layer.

series of values of θ, each corresponding to a different integral value of n, and a point X-ray beam will give rise to a series of points in a line perpendicular to the plane of the array. Provided the values of n can be assigned, the value of d may be calculated from measurement of θ for a given point [the angle between the transmitted and diffracted beams is 2θ (figure 10.18)]. The lamellar spacing of approximately 5·0 nm gives rise to rather low diffraction angles in the vertical plane, and special techniques are generally necessary to record diffractions from macromolecular systems such as membranes.

In addition to the lamellar diffractions, an ordered array of hydrocarbon chains in the bilayers of figure 10.18 will give rise to a diffraction pattern in the horizontal axis, parallel to the planes of the bilayers. An X-ray diffraction pattern from multilamellar arrays of egg phosphatidylcholine above its phase transition and at 14% water content is shown in figure 10.19. A series of sharp arcs cutting the vertical axis or meridian (perpendicular to the planes of the bilayers) corresponds to diffractions from lamellae 4·97 nm apart, while two diffuse 0·46 nm bands located about the horizontal axis (equator) arise from hydrocarbon chains oriented largely perpendicular to the lamellar planes. Despite this marked orientation of the hydrocarbon chains, there is a broad wide-angle diffraction band of low intensity cutting the vertical axis of the diffraction pattern corresponding to regions in which short segments of the hydrocarbon chains lie nearly parallel to the plane of the bilayer. This presumably corresponds to disorder of methyl terminal segments of the chain in the liquid crystalline state. Increased hydration of the bilayer leads to increasing disorder of the chains, and the diffraction pattern at 21% hydration shows the equatorial orientation of the wide-angle 0·46 nm band to be less marked than that at 14% hydration.

Similar diffraction patterns have been obtained from bilayers of synthetic phospholipids. As the temperature of dipalmitoyl phosphatidyl-choline is lowered through that of the phase transition, the diffuse 0·46 nm band changes to a sharp 0·42 nm diffraction characteristic of fully extended alkyl chains packed in a two-dimensional lattice. Direct measurements from the diffraction patterns indicate that the extended chains tilt away from a direction perpendicular to the plane of the bilayer with increasing hydration to a maximum of 60° (chapter 6, figure 6.14). This is confirmed by calculations of the average area occupied by each phospholipid in the plane of the bilayer. The thickness of the bilayer

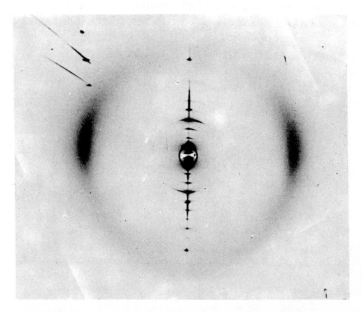

Figure 10.19. Diffraction pattern from oriented bilayers of egg phosphatidylcholine at 14% water content: from Y. K. Levine and M. H. F. Wilkins, *Nature New Biology* (1971), **230,** 69–72, by courtesy of Professor M. H. F. Wilkins, School of Biological Sciences, University of London, King's College.

increases by some 0·5 nm on transition to the gel phase as the alkyl chains assume fully extended conformations.

Measurements of the intensities of the diffracted beams can be used to calculate *electron density profiles* in the bilayers. Two such profiles from the egg phosphatidylcholine bilayer that gave the diffraction pattern of figure 10.19, are shown in figure 10.20. The peaks of electron density correspond to lipid head groups, and the central troughs of low electron density to the terminal methyl groups of the alkyl chains. At 14% hydration these methyl groups appear to be localized in the centre of the bilayer. Addition of water causes delocalization of the terminal methyl groups, and the central trough is broadened at 21% hydration. This corresponds to the increasing disorder of the terminal chain segments suggested by the decreased equatorial orientation of the 0·46 nm diffraction observed with increasing hydration.

Convenient interpretation of X-ray data depends upon the presence of a multilamellar array of centrosymmetric structures. The model lipid bilayers discussed above have themselves a centre of symmetry, but natural membranes tend to contain asymmetrically distributed protein, and a single bilayer is itself asymmetric. A very few biological membranes,

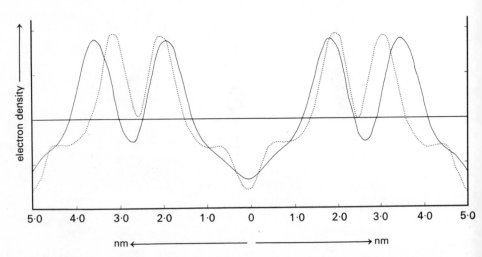

Figure 10.20. Electron density profiles of egg phosphatidylcholine bilayers at 14% water content (............) and 21% water content (————). Distances are measured perpendicular to the planes of the bilayers: from Y. K. Levine and M. H. F. Wilkins, *Nature New Biology* (1971), **230**, 69–72.

like myelin and retinal rod disc membranes, exist naturally as stacked pairs of apposed bilayers with a centre of symmetry within each two-membrane unit, and myelin in particular has been extensively studied by X-ray diffraction. An electron density profile for myelin from frog sciatic nerve is shown in figure 10.21.

The similarity to the pattern of the central lipid bilayer unit of figure 10.20 is obvious. Again the high-density peaks correspond to lipid head groups, but this time the inner (i.e. cytoplasmic side) peak is higher; a fact which is explained by a higher protein content on the cytoplasmic face of the bilayer. As in the model lipid system, the central trough

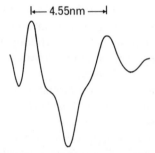

Figure 10.21. 'Best-fit' electron density profile for frog sciatic nerve myelin proposed by A. E. Blaurock (*Biophys. J.* (1976), **16**, 491–501) on the basis of X-ray diffraction data published by C. R. Worthington and T. J. McIntosh (*Biophys. J.* (1974), **14**, 703–729). The profile represents a single bilayer which is separated from its mirror image (on the left) by a 'cytoplasmic' gap. A very similar profile has been derived from independent data of D. L. D. Caspar and D. A. Kirschner (*Nature New Biology* (1971), **231**, 46–52).

reflects the presence of the terminal methyl groups of the hydrocarbon chains. The marked shoulders bordering this central trough occur in the same position as similar features in hydrated egg phosphatidylcholine-cholesterol (1:1) patterns and are believed to be characteristic of cholesterol [contrast figure 10.20 (21% water)].

Retinal rods would appear to be designed for X-ray diffraction studies containing, as they do, stacked pairs of membranes in flattened discs (figure 3.11). They are, however, relatively small and measurements present technical difficulties. Nevertheless, asymmetric electron density profiles similar to those of myelin have been obtained from frog outer rod segments and have been interpreted as reflecting a higher concentration of the major protein rhodopsin on the intradisc side of the lipid bilayer (but see pages 36 and 211).

Until relatively recently, X-ray diffraction studies had been confined to the natural multilamellar arrays of myelin, retinal rods and, to a lesser extent, chloroplasts. Techniques have, however, now been developed for sedimentation of stacks of flattened membrane vesicles, and electron density profiles of erythrocyte and of functional sarcoplasmic reticulum membranes have been reported which demonstrate the bilayer basis of these structures and allow conclusions to be drawn regarding the distribution of their protein components across the width of the bilayer. Similar bilayer patterns have been derived from studies on experimentally-stacked sheets of purple membrane from *Halobacterium halobium* and, rather more excitingly, analysis of strong diffraction peaks equatorially centred at 1·0 nm and axially at 0·5 nm and 0·15 nm has led to the concept that the major protein bacteriorhodopsin crosses the membrane as a complex of seven helices (figure 9.9), an arrangement supported by electron microscopy and diffraction.

X-ray diffraction has been applied to unoriented membrane dispersions but, although some information has been obtained from a range of membranes by this means, interpretation is subject to error and the most reliable information has generally been obtained from oriented multilamellar systems. Other diffraction techniques providing information complementary to that derived from X-ray diffraction have been applied to the study of membranes.

Neutron diffraction

Any beam of moving particles will display wave properties which depend upon the mass and velocity of the particles. Provided that the wavelength is suitable, the beam can, in principle, be used like X-rays to investigate structure. This is the case with neutrons. Neutrons of appropriate velocity are in fact emitted from atomic reactors, and the analysis of membranes by neutron scattering is at an early stage of development. Whereas X-rays are scattered by electrons, neutrons are scattered by the nuclei of the sample, and the use of neutrons has the advantages that electron-poor atoms such as hydrogen are readily detected and that scattering densities of membrane components vary much more than for X-rays. Specific deuteriation of membrane components can be used to enhance these differences, so aiding interpretation of the data from neutron scattering experiments, and the technique has to date been employed in limited studies of the model lipid bilayers, myelin and retinal rod outer segments.

Electron diffraction

Electrons also have wave properties and can be accelerated into a beam which is scattered on interaction with atomic electric fields. Whereas the technique of electron diffraction has been widely applied to the structural analysis of materials, however, it has been much less used in the analysis of biological membranes. Its major disadvantage is the damage that can be caused to the specimen by radiation, and also by the vacuum necessary to the technique. Nevertheless electron diffraction does have the advantage of being able to obtain scattering information from very small areas of specimen, and that imaging (i.e. conventional electron microscopy) can give complementary information from the same sample. The well-ordered array of bacteriorhodopsin in the purple membrane of *Halobacterium halobium* is particularly well suited to the combined approach of electron microscopy and diffraction which, as mentioned above, led to the three-dimensional protein structure shown in figure 9.9.

Raman scattering

In a Raman scattering experiment a fixed-frequency laser beam is focused into the sample and the scattered light is analysed for frequency and intensity in a spectrometer. Whereas most of the incident photons are scattered elastically with no frequency change (Rayleigh photons), a small percentage exchange energy with the vibrational states of the sample molecules and emerge with lower frequencies. The frequency changes of these Raman photons correspond to vibrational frequencies of molecules in the sample, and intensity variations at particular frequency shifts can be used to report on specific groups (e.g. C-H or C-C) within the sample.

Conformational changes associated with phase transitions in model membranes have been studied by these means and, while such applications of the technique are currently limited, it is possible that as more sophisticated developments occur they will increase. *Resonance Raman spectroscopy*, for example, involves excitation with a laser frequency that lies under a molecular electronic absorption band. This can lead to an increased probability of photon-molecule energy exchange and consequent intensity or resonance enhancement compared to the normal Raman effect.

Calorimetry

As the temperature of a pure phospholipid-water system is raised through that of its gel to liquid crystalline phase transition, the mobility of the hydrocarbon chains suddenly increases, and a corresponding abrupt rise in heat absorption occurs. This heat absorption can be monitored by calorimetric techniques, most commonly by *differential scanning calorimetry* (DSC). A differential scanning calorimeter consists of two cells, one of which contains sample and the other an inert reference material (e.g. water, when the sample is an aqueous lipid dispersion). The sample and reference are heated separately, so as to maintain zero temperature difference between the two cells; the differential heat flow necessary to maintain this relationship is recorded and plotted against sample temperature. At the phase transition of a lipid sample, the extra heat flow will be recorded as a peak, and the area under the peak can be related to the enthalpy change of the transition.

The DSC scans of dimyristoyl phosphatidylcholine and of dipalmitoyl phosphatidylcholine in excess water are shown in figures 10.22a and g respectively. In each case the large endothermic peak corresponds to the main gel-liquid crystalline phase transition of the hydrocarbon chains. The lower-temperature 'pre-transition' peaks are characteristic of pure phosphatidylcholines and do not occur in other synthetic lipids. The nature of the structural change associated with this pre-transition is not clear, although it has been suggested that the peak reflects a change of state of the polar phosphatidylcholine head groups. The temperature of the main transition of dipalmitoyl phosphatidylcholine can be seen (figure 10.22) to be higher than that of the pure dimyristoyl derivative and it is generally found that, in saturated phospholipids, the transition

Table 10.1. Transition temperatures for gel to liquid crystalline phase transitions in diacyl phosphatidylcholine, containing two identical chains, in excess water.

Fatty Acyl Chains	Transition Temperature (°C)
Dibehenoyl (22:0)	75
Distearoyl (18:0)	55
Dipalmitoyl (16:0)	41
Dimyristoyl (14:0)	24
Dilauroyl (12:0)	−2
Dioleoyl (18:1)	−22

temperature increases with increased fatty acid chain length. *Cis* unsaturated fatty acyl chains tend to reduce the transition temperature (Table 10.1).

The thermal behaviour of mixtures of phospholipids depends upon the similarity of the components. Thus dimyristoyl and dipalmitoyl phosphatidylcholines differ in just two carbons in their fatty-acyl chain lengths and, as can be seen in figure 10.22, mixtures of the phospholipids give

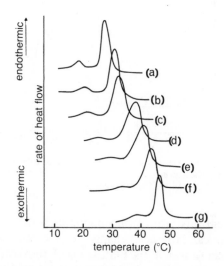

Figure 10.22. DSC heating curves for mixtures of dimyristoyl phosphatidylcholine and dipalmitoyl phosphatidylcholine in excess water. Contents of dimyristoyl phosphatidylcholine are (*a*) 100%; (*b*) 80%; (*c*) 70%; (*d*) 50%; (*e*) 30%; (*f*) 15%; (*g*) 0%: from D. Chapman *et al.*, *J. Biol. Chem.* (1974), **249**, 2512–2521.

DSC scans which indicate almost ideal mixing within the bilayer. The binary mixtures show broad transition peaks with maximum broadening at a ratio of 1:1, and the composite transition temperature approximates to a linear function of dipalmitoyl phosphatidylcholine content. Binary mixtures of dimyristoyl and distearoyl phosphatidylcholines differing by four carbons in their fatty-acyl chains give mixtures which are far from ideal, and DSC scans show very broad transition peaks of complex shape. On the other hand, when the components of a binary phospholipid mixture have widely different transition temperatures, as in the case of dilauroyl and distearoyl phosphatidylcholines, separate peaks can often be seen in DSC profiles (figure 10.23) suggesting fractional crystallization of the components within the bilayers. This phase separation does not, of course, occur when different fatty-acyl chains are in the same molecule (as in natural phospholipids), and 2-oleoyl-1-stearoyl-*sn*-glycero-3-phosphocholine shows a single sharp transition peak at a temperature between those of the distearoyl and dioleoyl derivatives.

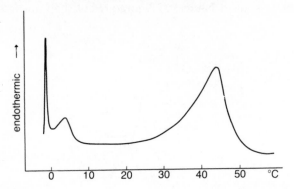

Figure 10.23. DSC heating curve for a 1:1 mixture of dilauroyl phosphatidylcholine and distearoyl phosphatidylcholine in excess water: from S. Mabrey and J. M. Sturtevant, *Proc. Nat. Acad. Sci.*, U.S. (1976), **73**, 3862–3866.

Two-component mixtures of phospholipids with different polar head groups but identical fatty-acyl chains do not necessarily show ideal mixing and, as can be seen in the DSC scans of dimyristoyl phosphatidylcholine and dimyristoyl phosphatidylethanolamine (figure 10.24), broad highly asymmetric composite peaks are obtained, the temperatures of which are not a linear function of composition.

The effect of adding cholesterol to a phospholipid is shown in the DSC scan of figure 10.25. At low cholesterol content the heat absorption can be resolved into two components, one sharp and one broad (figure 10.25*a*). The sharp peak corresponds to the transition temperature of the pure phospholipid, and this peak has completely disappeared at a cholesterol content greater than 20%. As can be seen in figure 10.25, the composite peak broadens and shifts to higher temperatures as the cholesterol content is increased, until above 50% cholesterol, when a transition is no longer detectable in the DSC profile.

The thermal behaviour so far described has been confined to that of pure lipids or binary mixtures dispersed in excess water (i.e. at least an equal weight of water), when the lipid exists in the form of multilam-

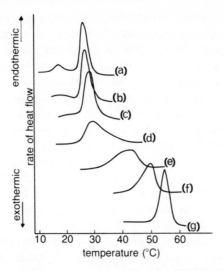

Figure 10.24. DSC heating curves for mixtures of dimyristoyl phosphatidylcholine and dimyristoyl phosphatidylethanolamine in excess water. Contents of dimyristoyl phosphatidylcholine are (*a*) 100%; (*b*) 95%; (*c*) 90%; (*d*) 80%; (*e*) 50%; (*f*) 20%; (*g*) 0%. Note the absence of a pretransition peak (characteristic of phosphatidylcholine) in pure dimyristoyl phosphatidylethanolamine (*g*): from D. Chapman *et al. J. Biol Chem* (1974), **249**, 2512–2521.

mellar bilayers (chapter 6). These conditions have been most studied because they are believed to form a simple model for natural membranes. At low water content (i.e. less than 20% by weight) lipid adopts conformations less relevant to natural membranes and the thermal behaviour is different (e.g. DSC scans no longer show sharp transition peaks).

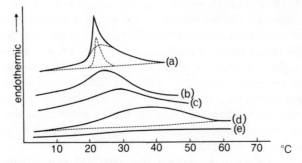

Figure 10.25. DSC heating curve for mixtures of cholesterol with dimyristoyl phosphatidylcholine in aqueous suspension. Contents of cholesterol are (*a*) 17%; (*b*) 20%; (*c*) 26%; (*d*) 33%; (*e*) 55%: from S. Mabrey *et al., Biophys. J.* (1977), **17**, 82a.

Sonicated lipid vesicles contain a single bilayer (page 104) and have been extensively used in transport and, to some extent, in magnetic resonance studies (page 241). DSC profiles of sonicated vesicles, however, show broad asymmetric peaks of low intensity, indicating differences between the hydrocarbon chain packing and freedom in sonicated and unsonicated vesicles. The present feeling is that, although sonicated vesicles with their small radius of curvature may well have physiological relevance in special situations, unsonicated preparations provide a better general experimental model for biological membrane structure.

Biological membranes are, of course, far more complex than the simple models discussed above. Not only do natural membranes contain a wide range of different lipid head groups and fatty-acyl chains which, as we have seen, will tend to broaden out phase transitions, but they also contain proteins and interact with metal ions, both of which can have similar effects. In addition, mammalian membranes usually have a high content of cholesterol and, not surprisingly, do not in general show lipid phase transitions on thermal analysis. Some evidence for broad lipid phase transitions at sub-physiological temperatures has been obtained from isolated mammalian systems such as rat liver microsomal membranes which, like intracellular membrane generally, are low in cholesterol, but most data on natural membranes come from bacteria.

Mycoplasmas are wall-less micro-organisms that require exogenous fatty acids and (apart from the *Acholeplasma* species) cholesterol for growth. They can accordingly be used to provide membranes of relatively defined composition for thermal and other studies, and in fact *Acholeplasma laidlawii* membranes were the first biological membranes in which a thermal phase transition was demonstrated. More recent studies have shown the presence in *A. laidlawii* of broad lipid phase transitions which can be artificially shifted over wide ranges of temperature by selective fatty-acid enrichment (saturated acids raise the transition temperature) but which are normally completed well below growth temperature. A second higher-temperature thermal transition has also been demonstrated in the membrane of *A. laidlawii*. This is attributed to protein denaturation and, in contrast to the lipid transition, is irreversible. It is of interest that neither denaturation nor enzymic cleavage of protein appears to affect the lipid transition, suggesting a lack of protein-lipid interaction in the membrane.

Thermal transitions have been demonstrated in a number of other bacterial membranes, particularly *E. coli*. *E. coli*, which has a complex cell envelope (page 7), normally contains a heterogeneous mixture of membrane lipids but mutants are available which have been used to follow the thermal consequences of selective fatty-acid enrichment, in much the same way as in *A. laidlawii*.

Calorimetry provides accurate information of overall membrane phase changes but cannot report in detail on the structural nature of these changes. It does, however, have the advantage of not perturbing the structure of the membrane and, as calorimeters become more sensitive, they may find an increasingly important role in providing data complementary to those obtained by physical techniques that report at a molecular level.

SUMMARY

1. ^{1}H, ^{13}C, ^{2}H, ^{31}P and ^{19}F nuclear magnetic resonance spectra have been used to report on the relative motion of the fatty-acyl chains and of the polar head groups of phospholipids in model bilayers. Suitably-located paramagnetic reference ions have also been used to modify n.m.r. signals and so to clarify the dispositions of resonating nuclei; n.m.r. spectra of biological membranes tend to be less easily interpreted and less information has been obtained about natural systems.

2. Nitroxide spin labels can be incorporated into lipid bilayers, and their electron spin resonance spectra can be used to report on the polarity of their environment and also on the detailed mobility of the probe and its host molecule in the bilayer. Both lateral and 'flip-flop' motion of the phospholipids in model bilayers have been measured using nitroxide probes.

3. Fluorescent probe molecules are sensitive to the polarity and to the viscosity of their environment and have been widely used to study both model and naturally-occurring bilayers. Specifically-located fluorescent probes have also been used to follow rapid changes in membrane state induced by processes such as electron-transport and changes in membrane potential. Membrane proteins generally contain tryptophan residues which can be used as intrinsic probes to report on conformational changes following binding of specific ligands. The lateral motions of components within the plane of a lipid bilayer have been followed by using the techniques of fluorescence photobleaching recovery, fluorescence correlation spectroscopy, and resonance energy transfer.

4. X-ray diffraction allows measurement of lamellar spacing, interchain distances, and electron densities in stacked arrays of lipid bilayers and also of biological membranes, provided that multilamellar arrays of centrosymmetric units can be obtained. Myelin and retinal rod disc membranes are particularly suitable in this respect, but artificial sedimentation techniques have been developed, by means of which other membranes can be arranged in stacks of flattened vesicles amenable to X-ray study. Neutron and electron diffraction techniques have been similarly, but to a much lesser extent, used in membrane structural studies. Raman and resonance Raman spectroscopy give information about the presence and vibrational states of particular chemical groupings within a bilayer.

5. Differential scanning calorimetry has been used to study phase transitions in lipid bilayers composed of pure phospholipids and of their mixtures. Cell membranes of the mycoplasmas have been studied because of the relative ease with which their membrane composition can be manipulated.

FURTHER READING

Chapter 1
De Robertis, E. D. P., Saez, F. A. and De Robertis, E. M. F. (1975) *Cell Biology,* 7th ed. (Chapters 1–6), W. B. Saunders, Philadelphia, U.S.A.

Chapter 2
Coleman, R. (1973) 'Membrane-bound enzymes and membrane ultrastructure' *Biochim. Biophys. Acta,* **300,** 1–30.
Finean, J. B., Coleman, R. and Michell, R. H. (1978) *Membranes and their Cellular Functions,* 2nd ed., Blackwell Scientific Publications, Oxford and London.

Chapter 3
Daemen, F. J. M. (1973) 'Vertebrate rod outer segment membranes' *Biochim. Biophys. Acta (Reviews on Biomembranes),* **300,** 255–288.
De Robertis, E. D. P., Saez, F. A. and De Robertis, E. M. F. (1975) *Cell Biology,* 7th ed. (Chapters 7–11), W. B. Saunders, Philadelphia, U.S.A.
Harris, J. R. (1979) 'The biochemistry and ultrastructure of the nuclear envelope' *Biochim. Biophys. Acta (Reviews on Biomembranes),* **515,** 56–103.
Matus, A. (1978) 'Synaptic membranes and junctions from brain' in *Methods in Membrane Biology,* vol. 9 (ed. E. D. Korn), Plenum Press, New York and London, 203–236.
Munn, E. A. (1974) *The Structure of Mitochondria,* Academic Press, London.
Novikoff, A. B. and Holtzman, E. (1976) *Cells and Organelles,* 2nd ed., Holt, Rinehart and Winston, London and New York.
Staehelin, L. A. (1976) 'Reversible particle movements associated with unstacking and restacking of chloroplast membranes *in vitro*' *J. Cell. Biol.,* **71,** 136–158.
Staehelin, L. A. and Hull, B. A. (1978) 'Junctions between living cells' *Scientific American,* **238,** 141–152.

Chapter 4
Baker, R. M. and Ling, V. (1978) 'Membrane mutants of mammalian cells in culture' in *Methods in Membrane Biology,* vol. 9 (ed. E. D. Korn), Plenum Press, New York and London, 337–384.
De Pierre, J. W. and Karnovsky, M. L. (1973) 'Plasma membranes of mammalian cells. A review of methods for their characterisation and isolation' *J. Cell Biol.,* **56,** 275–303.
Di Rienzo, J. M., Nakamura, K. and Inouye, M. (1978) 'The outer membrane proteins of Gram-negative bacteria: biosynthesis, assembly and functions' *Ann. Rev. Biochem.,* **47,** 481–532.
Lenard, J. (1978) 'Virus envelopes and plasma membranes' *Ann. Rev. Biophys. Bioeng.,* **7,** 139–165.
Patton, S. and Keenan, T. W. (1975) 'The milk fat globule membrane' *Biochim. Biophys. Acta (Reviews on Biomembranes),* **415,** 273–309.

Razin, S. and Rottem, S. (1978) 'Cholesterol in membranes: studies with mycoplasmas' *TIBS*, **3**, 51–55.

Salton, M. R. J. and Owen, P. (1976) 'Bacterial membrane structure' *Ann. Rev. Microbiol.*, **30**, 451–482.

Thompson, G. A. and Nozawa, Y. (1978) 'Tetrahymena: A system for studying dynamic membrane alterations within the eukaryotic cell' *Biochim. Biophys. Acta (Reviews on Biomembranes)*, **472**, 55–92.

Wallach, D. F. H. and Lin, P. S. (1973) 'A critical evaluation of plasma membrane fractionation' *Biochim. Biophys. Acta (Reviews on Biomembranes)*, **300**, 211–254.

Chapter 5

Ansell, G. B., Hawthorne, J. N. and Dawson, R. M. C. (eds.) (1973) *Form and Function of Phospholipids*, 2nd ed., Elsevier, Amsterdam.

Brennan, P. J. and Lösel, D. M. (1978) 'Physiology of fungal lipids: selected topics' *Adv. Microb. Physiol.*, **17**, 47–179.

Demel, R. A. and de Kruyff, B. (1976) 'The function of sterols in membranes' *Biochim. Biophys. Acta*, **457**, 109–132.

Di Rienzo, J. M., Nakamura, K. and Inouye, M. (1978) 'The outer membrane proteins of Gram-negative bacteria: biosynthesis, assembly, and functions' *Ann. Rev. Biochem.*, **47**, 481–532.

Finnerty, W. R. (1978) 'Physiology and biochemistry of bacterial phospholipid metabolism' *Adv. Microb. Physiol.*, **18**, 177–233.

Furthmayr, H. (1978) 'Glycophorins A, B and C: A Family of Sialoglycoproteins. Isolation and Preliminary Characterization of Trypsin Derived Peptides' *J. Supramol. Struct.*, **9**, 79–95.

IUPAC—IUB Commission on Biochemical Nomenclature (1978) 'The nomenclature of lipids (Recommendations 1976)' *Biochem. J.*, **171**, 21–35.

Lux, S. E. (1979) 'Dissecting the red cell membrane skeleton' *Nature*, **281**, 426–429

Maddy, A. H. (ed.) (1976) *Biochemical Analysis of Membranes*, Chapman and Hall, London.

Marchesi, V. T. Furthmayr, H. and Tomita, M. (1976) 'The red cell membrane' *Ann. Rev. Biochem.*, **45**, 667–671.

Moreau, F., Dupont, J. and Lance, C. (1974) 'Phospholipid and fatty acid composition of outer and inner membranes of plant mitochondria' *Biochim. Biophys. Acta*, **345**, 294–304.

Philipp, E-I, Franke, W. W., Keenan, T. W., Stadler, J. and Jarasch, E-D. (1976) 'Characterization of nuclear membranes and endoplasmic reticulum isolated from plant tissue' *J. Cell Biol.*, **68**, 11–29.

Phutrakul, S. and Jones, M. N. (1979) 'The effect of the incorporation of erythrocyte membrane extracts on the permeability of bilayer lipid membranes and the identification of the monosaccharide transport proteins' *Biochim. Biophys. Acta*, **550**, 188–200.

Pinder, J. C., Ungewickell, E., Bray, D. and Gratzer, W. B. (1978) 'The spectrin-actin complex and erythrocyte shape' *J. Supramol. Struct.*, **8**, 439–445.

Rouser, G., Nelson, G. J., Fleischer, S. and Simon G. (1968) 'Lipid composition of animal cell membranes, organelles and organs' in *Biological Membranes, Physical Fact and Function* (ed. D. Chapman), Academic Press, London, 5–69.

Rubin, R. W. and Milikowski, C. (1978) 'Over two hundred polypeptides resolved from the human erythrocyte membrane' *Biochim. Biophys. Acta*, **509**, 100–110.

Schwertner, H. A. and Biale, J. B. (1973) 'Lipid composition of plant mitochondria and chloroplasts' *J. Lipid Res.*, **14**, 235–242.

Shaw, N. (1974) 'Lipid composition as a guide to the classification of bacteria' *Adv. App. Microbiol.*, **17**, 63–108.

Shaw, N. (1975) 'Bacterial glycolipids and glycophospholipids' *Adv. Microb. Physiol.*, **12**, 141–167.

Steck, T. L. (1978) 'The Band 3 Protein of the Human Red Cell Membrane: A Review' *J. Supramol. Struct.*, **8**, 311–324.

Tanner, M. J. A. (1978) 'Erythrocyte glycoproteins' in *Current Topics in Membranes and Transport*, vol. 11 (ed. F. Bronner *et al.*), Academic Press, New York, 279–325.

Tomita, M. Furthmayr, H. and Marchesi, V. T. (1978) 'Primary structure of human erythrocyte glycophorin A. Isolation and characterization of peptides and complete amino acid sequence' *Biochemistry*, **17**, 4756–4770.

Chapter 6

Clarke, M. and Spudich, J. A. (1977) 'Nonmuscle contractile proteins: the role of actin and myosin in cell motility and shape determination' *Ann. Rev. Biochem.*, **46**, 797–822.

Fowler, V. and Bennett, V. (1978) 'Association of spectrin with its membrane attachment site restricts lateral mobility of human erythrocyte integral membrane proteins' *J. Supramol. Struct.*, **8**, 215–221.

Goodman, S. R. and Branton, D. (1978) 'Spectrin binding and the control of membrane protein mobility' *J. Supramol. Struct.*, **8**, 455–463.

Lazarides, E. and Revel, J. P. (1979) 'The molecular basis of cell movement' *Scientific American*, **240**, 88–100.

Lee, A. G. (1977) 'Lipid phase transitions and phase diagrams. II Mixtures involving lipids' *Biochim. Biophys. Acta*, **472**, 285–344.

Lodish, H. F. and Rothman, J. E. (1979) 'The assembly of cell membranes' *Scientific American*, **240**, 35–53.

Morré, D. J., Kartenbeck, J. and Franke, W. W. (1979) 'Membrane flow and interconversions among endomembranes' *Biochim. Biophys. Acta*, **559**, 71–152.

Nicolson, G. L. (1976) 'Transmembrane control of the receptors on normal and tumour cells. I Cytoplasmic influence over cell surface components' *Biochim. Biophys. Acta*, **457**, 57–108.

Rothman, J. E. and Lenard, J. (1977) 'Membrane asymmetry' *Science*, **195**, 743–753.

Schlessinger, J., Axelrod, D., Koppel, D. E., Webb, W. W. and Elson, E. L. (1977) 'Lateral transport of a lipid probe and labelled proteins on a cell membrane' *Science*, **195**, 307–309.

Wirtz, K. W. A. and van Deenan, L. L. M. (1977) 'Phospholipid-exchange proteins: a new class of intracellular lipoproteins' *Trends in Biochem. Sci.*, 49–51.

Verkleij, A. J. and Ververgaert, P. H. J. Th. (1978) 'Freeze-fracture morphology of biological membranes' *Biochim. Biophys. Acta*, **515**, 303–327.

Chapter 7

General

Cook, G. M. W. and Stoddart, R. W. (1973) *Surface Carbohydrates of the Eukaryotic Cell*, Academic Press, London.

Hughes, R. C. (1976) *Membrane Glycoproteins—A Review of Structure and Function*, Butterworths, London.

Purification and analysis of membrane carbohydrate

Cook, G. M. W. (1976) 'Techniques for the analysis of membrane carbohydrates' in *Biochemical Analysis of Membranes* (ed. A. H. Maddy), Chapman and Hall, London, 283–351.

Structure and biosynthesis of membrane glycoproteins

Chen, W. W. and Lennarz, W. J. (1978) 'Enzymic excision of glucosyl units linked to the oligosaccharide chains of glycoproteins' *J. Biol. Chem.*, **253**, 5780–5785.

Parodi, A. J. and Leloir, L. F. (1979) 'The role of lipid intermediates in the glycosylation of proteins in the eucaryotic cell' *Biochim. Biophys. Acta*, **559**, 1–37.

Robbins, P. W., Hubbard, S. C., Turco, S. J. and Wirth, D. F. (1977) 'Proposal for a common oligosaccharide intermediate in the synthesis of membrane glycoproteins' *Cell*, **12**, 893–900.

Staneloni, R. J. and Leloir, L. F. (1979) 'The biosynthetic pathway of the asparagine-linked oligosaccharides of glycoproteins' *Trends in Biochem. Sci.*, 65–67.

Sturgess, J., Moscarello, M. and Schachter, H. (1978) 'The structure and biosynthesis of membrane glycoproteins' in *Current Topics in Membranes and Transport*, vol. **11** (ed. F. Bronner *et al.*), Academic Press, New York, 15–104.

Tabas, I. and Kornfeld, S. (1978) 'The synthesis of complex-type oligosaccharides. III' *J. Biol. Chem.*, **253**, 7779–7786.

Waechter, C. J. and Lennarz, W. J. (1976) 'The role of polyprenol-linked sugars in glycoprotein synthesis' *Ann. Rev. Biochem.*, **45**, 95–112.

Blood group specific antigens (ABO)

Hakomori, S-i, Watanabe, K. and Laine, R. A. (1977) 'Glycosphingolipids with blood group A, H and I activity and their changes associated with ontogenesis and oncogenesis' *Pure App. Chem.*, **49**, 1215–1227.

Hanfland, P. (1975) 'Characterization of B and H blood group active glycosphingolipids from human B erythrocyte membranes' *Chem. Phys. Lipids*, **15**, 105–124.

Gardas, A. (1978) 'Structure of an (A-blood group)—active glycolipid isolated from human erythrocytes' *Eur. J. Biochem.*, **89**, 471–473.

Watkins, W. M. (1972) 'Blood group specific substances' in *Glycoproteins, their Composition, Structure and Function*, Part B (ed. A. Gottschalk), Elsevier, Amsterdam, 830–891.

Blood group specific antigens (MNSs)

Blumenfeld, O. O. and Admany, A. M. (1978) 'Structural polymorphism within the amino-terminal region of MM, NN and MN glycoproteins (glycophorins) of the human erythrocyte membrane' *Proc. Natl. Acad. Sci.*, **75**, 2727–2731.

Dahr, W., Uhlenbruck, G., Janssen, E. and Schmalisch, R. (1977) 'Different *N*-terminal amino acids in the MN-glycoprotein from MM and NN erythrocytes' *Hum. Genet.*, **35**, 335–343.

Furthmayr, H. (1978) 'Structural comparison of glycophorins and immunochemical analysis of genetic variants' *Nature*, **271**, 519–524.

Lisowska, E. and Wasniowska, K. (1978) 'Immunochemical characterization of cyanogen bromide degradation products of M and N blood group glycopeptides' *Eur. J. Biochem.*, **88**, 247–252.

Sadler, J. E., Paulson, J. C. and Hill, R. L. (1979) 'The role of sialic acid in the expression of human MN blood group antigens' *J. Biol. Chem.*, **254**, 2112–2119.

Springer, G. F. and Yang, H. J. (1977) 'Isolation and partial characterization of blood group M- and N- specific glycopeptides and oligosaccharides from human erythrocytes' *Immunochemistry*, **14**, 497–502.

Tomita, M., Furthmayr, H. and Marchesi, V. T. (1978) 'Primary structure of human erythrocyte glycophorin A. Isolation and characterization of peptides and complete amino acid sequence' *Biochemistry*, **17**, 4756–4770.

Histocompatibility and differentiation antigens

Barnstaple, C. J., Jones, E. A. and Crumpton, M. J. (1978) 'Isolation, structure and genetics of HLA-A, -B, -C and -DRw (Ia) antigens' *Br. Med. Bull.*, **34**, 241–246.

Bodmer, W. F. (1978) 'The HLA system: introduction' *Br. Med. Bull.*, **34**, 213–216.

Cunningham, B. A. (1977) 'The structure and function of histocompatibility antigens' *Scientific American*, **237**, 96–107.

Letarte, M. (1978) 'Glycoprotein antigens of murine lymphocytes' in *Current Topics in Membranes and Transport,* vol. **11** (ed. F. Bronner *et al.*), Academic Press, New York, 463–512.

McKenzie, I. F. C., Clarke, A. and Parish, C. R. (1977) 'Ia antigenic specificities are oligosaccharide in nature: Hapten inhibition studies' *J. Exp. Med.*, **145**, 1039–1053.

Snary, D., Barnstaple, C., Bodmer, W. F., Goodfellow, P. and Crumpton, M. J. (1977) 'Human Ia antigens—purification and molecular structure' *Cold Spring Harbor Symp. Quant. Biol.*, **41**, 379–386.

Springer, T. A., Kaufman, J. F., Terhorst, C., and Strominger, J. L. (1977) 'Purification and structural characterization of human HLA-linked B-cell antigens' *Nature*, **268**, 213–218.

Williams, A. F., McMaster, R., Standring, R. and Sunderland, C. A. (1978) 'Differentiation antigens and glycoproteins of lymphocytes' *Trends in Biochem. Sci.*, 272–274.

Receptors for bacterial toxins and glycoprotein hormones

Draper, R. K., Chin, D. and Simon, M. I. (1978) 'Diphtheria toxin has the properties of a lectin' *Proc. Natl. Acad. Sci.*, USA, **75**, 261–265.

Kohn, L. D. (1978) 'Relationships in the structure and function of receptors for glycoprotein hormones, bacterial toxins and interferon' in *Receptors and Recognition*, vol. **A5** (ed. P. Cuatrecasas and M. L. Greaves), Chapman and Hall, London, 133–212.

Pappenheimer, Jr., A. M. (1978) 'Diphtheria: molecular biology of an infectious process' *Trends in Biochem. Sci.*, N220–224.

van Heyningen, W. E. (1974) 'Gangliosides as membrane receptors for tetanus toxin, cholera toxin and serotonin' *Nature*, **249**, 415–417.

Liver receptors for desialylated plasma glycoproteins, erythrocytes and lymphocytes

Aminoff, D., Bell, W. C. and Vorder Bruegge, W. G. (1978) 'Cell surface carbohydrate recognition and the viability of erythrocytes in circulation' in *Cell Surface Carbohydrates and Biological Recognition* (ed. V. J. Marchesi *et al.*), Alan R. Liss, Inc., New York, 569–581.

Ashwell, G. and Morell, A. G. (1977) 'Membrane glycoproteins and recognition phenomena' *Trends in Biochem. Sci.*, 76–78.

Gesner, B. M., Woodruff, J. J. and McCluskey, R. T. (1969) 'An autoradiographic study of the effect of neuraminidase or trypsin on transfused lymphocytes' *Amer. J. Pathol.*, **57**, 215–224.

Paulson, J. C., Hill, R. L., Tanabe, T. and Ashwell, G. (1977) 'Reactivation of asialo-rabbit liver binding protein by resialylation with β-D-galactoside $\alpha2 \rightarrow 6$ sialyltransferase' *J. Biol. Chem.*, **252**, 8624–8628.

Receptor sites for viruses and bacteria

Gottschalk, A., Belyavin, G. and Biddle, F. (1972) 'Glycoproteins as influenza virus haemagglutinin inhibitors and as cellular virus receptors' in *Glycoproteins, Their Composition, Structure and Function Part B* (ed. A. Gottschalk), Elsevier, Amsterdam, 1082–1096.

Helenius, A., Morein, B., Fries, E., Simons, K., Robinson, P., Schirrmacher, V., Terhorst, C. and Strominger, J. L. (1978) 'Human (HLA-A and HLA-B) and murine (H-2K and H-2D) histocompatibility antigens are cell surface receptors for Semliki Forest virus' *Proc. Natl. Acad. Sci.*, USA, **75**, 3846–3850.

Ofek, I., Beachey, E. H. and Sharon, N. (1978) 'Surface sugars of animal cells as determinants of recognition in bacterial adherence' *Trends in Biochem. Sci.*, 159–160.

Waterfield, M. D., Espelie, K. and Elder, K. (1979) 'Structure of the haemagglutinin of influenza virus' *Br. Med. Bull.*, **35**, 57–63.

Cell-cell adhesion
Frazier, W. A. (1978) 'The role of cell surface components in the morphogenesis of the cellular slime molds' *Trends in Biochem. Sci.*, 130–133.
Glaser, L. (1978) 'Cell-cell adhesion studies with embryonal and cultured cells' *Rev. Physiol. Pharmacol.*, **83**, 89–122.
Hynes, R. O. and Destree, A. T. (1978) 'Relationships between fibronectin (LETS protein) and actin' *Cell*, **15**, 875–886.
Pena, S. D. J. and Hughes, R. C. (1978) 'Fibronectin-plasma membrane interaction in the adhesion and spreading of hamster fibroblasts' *Nature*, **276**, 80–83.
Rees, D. A., Lloyd, C. W. and Thom, D. (1977) 'Control of grip and stick in cell adhesion through lateral relationships of membrane glycoproteins' *Nature*, **267**, 124–128.
Thom, D., Powell, A. J. and Rees, D. A. (1979) 'Mechanisms of cellular adhesion IV. Role of serum glycoproteins in fibroblast spreading on glass' *J. Cell. Sci.*, **35**, 281–305.
Yamada, K. M. and Olden, K. (1978) 'Fibronectins—adhesive glycoproteins of cell surface and blood' *Nature*, **275**, 179–184.

Surface changes in cancer cells
Bramwell, M. E. and Harris, H. (1979) 'Some further information about the abnormal membrane glycoprotein associated with malignancy' *Proc. R. Soc. London*, B, **203**, 93–99.
Hakomori, S-i. (1975) 'Structures and organization of cell surface glycolipids. Dependency on cell growth and malignant transformation' *Biochim. Biophys. Acta*, **417**, 55–89.
Nicolson, G. L. (1976) 'Trans-membrane control of the receptors on normal and tumour cells. II. Surface changes associated with transformation and malignancy' *Biochim. Biophys. Acta*, **458**, 1–72.
Nicolson, G. L. (1979) 'Cancer metastasis' *Scientific American*, **240**, 50–60.
Old, L. J. (1977) 'Cancer immunology' *Scientific American*, **236**, 62–79.

Chapter 8
Deves, R. and Krupka, R. M. (1978) 'Testing transport models with substrates and reversible inhibitors' *Biochim. Biophys. Acta*, **513**, 156–172.
Harold, F. M. (1977) 'Membranes and energy transduction in bacteria' *Current Topics in Bio-energetics*, **6**, 83–149.
Lauger, P. (1979) 'A channel mechanism for electrogenic ion pumps' *Biochim. Biophys. Acta*, **552**, 143–161.
Limas, C. J. (1978) 'Calcium transport ATPase of cardiac sarcoplasmic reticulum in experimental hyperthyroidism' *Am. J. Physiol.*, **235**, H745-H752.
Naftalin, R. J. and Holman, G. D. (1977) 'Transport of sugars in human red cells' in *Membrane Transport in Red Cells* (eds. J. C. Ellory and V. L. Lew), Academic Press, London, pp. 257–300.
Postma, P. W. and Roseman, S. (1976) 'The bacterial phosphoenolpyruvate: sugar phosphotransferase system' *Biochim. Biophys. Acta*, **457**, 213–257.
Pressman, B. C. (1976) 'Biological applications of ionophores' *Ann. Rev. Biochem.*, **45**, 501–530.
Phutrakul, S. and Jones, M. N. (1979) 'The permeability of bilayer lipid membranes on the incorporation of erythrocyte membrane extracts and the identification of the monosaccharide transport proteins' *Biochim. Biophys Acta*, **550**, 188–200.
Rosen, B. P. (1978) *Bacterial Transport* (Chapters 4 and 12), Plenum Press, London and New York.

Semenza, E., Semenza, G. and Carafoli, E. (eds.) (1977) *Biochemistry of Membrane Transport*, FEBS Symposium No. 42, Springer-Verlag, Berlin.

Simoni, R. D. and Postma, P. W. (1975) 'The energetics of bacterial active transport' *Ann. Rev. Biochem.*, **44**, 524–554.

Wilson, D. B. (1978) 'Cellular transport mechanisms' *Ann. Rev. Biochem.*, **47**, 933–965.

Chapter 9

Avron, M. (1977) 'Energy transduction in chloroplasts' *Ann. Rev. Biochem.*, **46**, 143–155.

Catt, K. J., Harwood, J. P., Aguilera, G. and Dufau, M. L. (1979) 'Hormonal regulation of peptide receptors and target cell responses' *Nature*, **280**, 109–116.

Fain, J. N. (1978) 'Hormones, membranes and cyclic nucleotides' in *Receptors and Recognition*, Series A, No. 6 (eds. P. Cuatrecasas and M. F. Greaves), Chapman and Hall, London, pp. 1–61.

Gill, M. D. and Meren, R. (1978) 'ADP-ribosylation of membrane proteins catalysed by cholera toxin: Basis of the activation of adenylate cyclase' *Proc. Natl. Acad. Sci.*, USA, **75**, 3050–3054.

Green, D. E. (1977) 'Mechanism of action of uncouplers on mitochondrial energy coupling' *Trends in Biochem. Sci.*, **2**, 113–116.

Greengard, P. (1976) 'Possible role for cyclic nucleotides and phosphorylated membrane proteins in postsynaptic actions of neurotransmitters' *Nature*, **260**, 101–107.

Hatefi, Y. and Djavadi-Ohaniance, L. (eds.) (1976) *The Structural Basis of Membrane Function*, Academic Press, London.

Heidmann, T. and Changeux, J-P. (1978) 'Structural and functional properties of the acetylcholine receptor protein in its purified and membrane bound state' *Ann. Rev. Biochem.*, **47**, 317–57.

Henderson, R. (1977) 'The purple membrane from *Halobacterium halobium*' *Ann. Rev. Biophys. Bioeng.*, **6**, 87–109.

Keynes, R. D. (1979) 'Ion channels in the nerve-cell membrane' *Scientific American*, **240**, 98–107.

Kozlov, I. A. and Skulachev, V. P. (1977) 'H^+-adenosine triphosphatase and membrane energy coupling' *Biochim. Biophys. Acta*, **363**, 28–89.

O'Brien, P. J. (1978) 'Rhodopsin: a light sensitive membrane glycoprotein' in *Receptors and Recognition*, Series A, No. 6 (eds. P. Cuatrecasas and M. F. Greaves), Chapman and Hall, London, pp. 107–150.

Rojas, E. and Bergman, C. (1977) 'Gating currents: molecular transitions associated with the activation of sodium channels in nerve' *Trends in Biochem. Sci.*, **2**, 6–9.

Chapter 10
General

Anderson, H. C. (1978) 'Probes of membrane structure' *Ann. Rev. Biochem.*, **47**, 359–383.

Nuclear Magnetic Resonance

Bergelson, L. D. (1978) 'Paramagnetic hydrophilic probes in NMR investigations of membrane systems' in *Methods in Membrane Biology* (ed. E. D. Korn), vol. **9**, Plenum Press, London, 275–335.

Cullis, P. R. and McLaughlin, A. C. (1977) 'Phosphorus nuclear magnetic resonance studies of model and biological membranes' *Trends in Biochem. Sci.*, 196–199.

Lee, A. G. (1978) 'Fluorescence and NMR studies of membranes' in *Receptors and Recognition*, vol. **A5** (ed. P. Cuatrecasas and M. F. Greaves), Chapman and Hall, London, 89–131.

Mantsch, H. H., Saito, H. and Smith, I. C. P. (1977) 'Deuterium magnetic resonance, applications in chemistry, physics and biology' *Prog. NMR Spectroscopy*, **11**, 211–271.

Seelig, J. (1977) 'Deuterium magnetic resonance: theory and applications to lipid membranes' *Quart. Rev. Biophys.*, **10**, 353–418.
Seelig, J. (1978) '^{31}P Nuclear magnetic resonance and the head group structure of phospholipids in membranes' *Biochim. Biophys. Acta*, **515**, 105–140.

Electron spin resonance
Berliner, L. J. (ed.) (1976) *Spin Labelling, Theory and Applications*, Academic Press, New York.
Gaffney, B. J. and Chen, S-C. (1977) 'Spin-label studies of membranes' in *Methods in Membrane Biology*, vol. **8** (ed. E. D. Korn), Plenum Press, London, 291–358.
Schreier, S., Polnaszek, C. F. and Smith, I. C. P. (1978) 'Spin labels in membranes. Problems in practice' *Biochim. Biophys. Acta*, **515**, 375–436.

Fluorescence
Lee, A. G. (1978) 'Fluorescence and NMR studies of membranes' in *Receptors and Recognition*, vol. **A5** (ed. P. Cuatrecasas and M. F. Greaves), Chapman and Hall, London, 81–131.
Radda, G. K. (1975) 'Fluorescent probes in membrane studies' in *Methods in Membrane Biology*, vol. **4** (ed. E. D. Korn), Plenum Press, New York, 97–188.
Shinitzsky, M. and Barenholz, Y. (1978) 'Fluidity parameters of lipid regions determined by fluorescence polarization' *Biochim. Biophys. Acta*, **515**, 367–394.

Diffraction methods
Carey, P. R. (1978) 'Resonance Raman spectroscopy in biochemistry and biology' *Quart. Rev. Biophys.*, **11**, 309–370.
Henderson, R. and Unwin, P. N. T. (1975) 'Three dimensional model of purple membrane obtained by electron microscopy' *Nature*, **257**, 28–32.
Herbette, L., Marguadt, J., Scarpa, A. and Blasie, J. K. (1977) 'A direct analysis of lamellar X-ray diffraction from hydrated orientated multilayers of fully functional sarcoplasmic reticulum' *Biophys. J.*, **20**, 245–272.
Hui, S. W. (1977) 'Electron diffraction studies of membranes' *Biochim. Biophys. Acta*, **472**, 345–371.
Pape, E. H., Klott, K. and Kreutz, W. (1977) 'The determination of the electron density profile of the human erythrocyte membrane by small-angle X-ray diffraction' *Biophys. J.*, **19**, 141–160.
Schoenborn, B. P. (1976) 'Neutron scattering for the analysis of membranes' *Biochim. Biophys. Acta*, **457**, 41–55.
Worthington, C. R. (1976) 'X-ray studies on membranes' in *The Enzymes of Biological Membranes* (ed. A. Martonosi), Plenum Press, 1–29.

Calorimetry
Mabrey, S. and Sturtevant, J. M. (1978) 'High sensitivity differential scanning calorimetry in the study of biomembranes and related model systems' in *Methods in Membrane Biology*, vol. **9** (ed. E. D. Korn), Plenum Press, New York, 237–274.
Melchior, D. L. and Steim, J. M. (1976) 'Thermotropic transitions in biomembranes' *Ann. Rev. Biophys. Bioeng.*, **5**, 205–238.

INDEX

boundary lipid 117
brain membranes 138, 153

C

Ca^{++}-ATPase 116, 183–185, 213–214, 260
Ca^{++} ions 27, 39
  and action potential in *Paramecium* 228
  and human receptors 232–233
  and muscle contraction 183–185
  and rhodopsin 212–214
calmodulin 233
calorimetry 268–273
cancer cells 162–164
Candida utilis 247
capping 119
cardiac glycosides 180–181
cardiolipin, see bis (phosphatidyl) glycerol
catecholamines 229
cell-cell adhesion 159–162
cell disruption 42–44
cell electrophoresis 133–134
cell junctions 27–28, *31*, see also junctional
    complexes
cell membrane 1–4
cell surface protein, see fibronectins
cell wall 1, 130
cephalin 70, see also phosphatidylethanolamine
ceramide 74, *75*, 143
cerebronic acid 78
cerebrosides 77–78, *79*
ceruloplasmin 156
chaotropic agents 90, 95, 135
chemical shift 240–241
chemical shift anisotropy 240, 245
chemiosmotic theory 186–188, 203
chlorophyll 23, 202
chloroplast 6, 23–27, *26*, *29*, 39, 112
cholera toxin 154–155, 193, 232, 259–260
cholesterol 81, *84*–85, 87, 88, 104, 123, 246, 254,
    265
cholesterol and membrane fluidity 58, 84,
    106–107, 116, 257, 271
chromaffin granule membranes 246
Cig 161, see cold insoluble globulin
colchicine 119
cold insoluble globulin, see fibronectins
colicins 173
collagen 121, 138, 161, 162, 163
compartmentalization 9–10
complement 149
concanavalin A 56, 113, 132, 135, 164, 261
coupling ATPase 23, *24*, 190, 195–206, 211
cristae 21, *22*, 23

CSP 161, see cell surface protein
cyanobacteria 8, 25
cyclic AMP 155
cyclic AMP and hormone receptors 229–236
cyclic AMP and synaptic receptors 226–227
cyclic AMP phosphodiesterase 230
cyclic GMP 230, 232–234
cytochalasin B 119
cytochemical techniques 130–131
cytochromes 92, 204–205
cytoskeleton 12, 119–121, 126, 162–163

D

Davson-Danielli-Robertson model 102–108
desmosomes 27, *31*
detergents 90, 91, 98, 135
deuterium 243–245, 247
diacylglycerol 86
diarrhoea 154
Dictyostelium discoideum 160
dicyclohexylcarbodiimide 13, 197
differential scanning calorimetry 268–273
differentiation antigens 153–154
diffraction methods 261–268
diffusion 166–167
digitonin 53
3,3-dihexyloxacarbocyanine 259
dihydrophytyl alcohol 74
diphtheria toxin 155
discoidins 160
dolichol 140–143
Donnan equilibrium 178
down regulation 235–236
DPH fluorescent probe *257*, 258
DRw antigens 152–153

E

electric eel 160, 180, 181, 224
electrochemical concentration gradient of
    protons, $\Delta\tilde{\mu}_{H+}$ 186–187, 197–199, 211
electrochemical concentration gradient of
    sodium ions, $\Delta\tilde{\mu}_{Na+}$ 190, 193
electrochemical gradients 11–12, 186–187
electron density profiles 264–266
electron diffraction 210, 267
electron spin resonance 104, 115–117, 247–255
electron transport 23, 200–206, 259
electronic energy states 255
Electrophorus electricus, see electric eel
electroplaques 218–220
endoglycosidase, see glycosidase
endoplasmic reticulum 5, 7, 87, 88, 125–128, 140
enveloped viruses, see viral membranes